Coastal and Estuarine Studies

Coastal and Estuarine Studies

formerly Lecture Notes on Coastal and Estuarine Studies

37

C.M. Lalli (Ed.)

Enclosed Experimental Marine Ecosystems: A Review and Recommendations

A Contribution of the Scientific Committee on Oceanic Research Working Group 85

Springer-Verlag
New York Berlin Heidelberg London Paris Tokyo Hong Kong

Editor

C.M. Lalli
The University of British Columbia, Department of Zoology
Vancouver, B.C. Canada V6T 2A9

ISBN-13: 978-1-4684-6403-0 e-ISBN-13: 978-1-4684-6401-6
DOI: 10.1007/978-1-4684-6401-6

Softcover reprint of the hardcover 1st edition 1990

2837/3140-543210 – Printed on acid-free paper

PREFACE

The application of mesocosms, defined in this report as artificial experimental enclosures ranging in size from 1 m^3 to 10 m^3, to address various problems in the marine sciences has been a relatively recent development. The application of the technology was dictated by the realization that many important ocean processes and interactions cannot be fully understood from observations in the natural environment or in smaller enclosures. Such studies involve, for example, determining the interactions between, and energy transfer from, one trophic level to another, the biogeochemical cycling of elements and compounds, etc. These and similar interactions and rate processes cannot normally be established in situations (nature) where the detection and quantification of rate processes are confused by advection and/or the inability to study the same populations over time. In the case of microcosms, mixed populations of primary producers, consumers, and carnivores cannot be maintained, in balance, for a sufficient length of time to determine normal interactions between the various components of these trophic levels.

This report, prepared by SCOR Working Group 85, critically examines past applications of mesocosms to ocean research, though there is no attempt to comprehensively review all literature relevant to the subject. Further, the report outlines some important advances emanating from their use and provides recommendations for future applications. It constitutes the first of two reports from the Working Group (see Introduction).

As advances in mesocosm technology have developed (and continue to be developed), a great diversity in the structure/design of such facilities can be expected. This is so because, as pointed out in several contributions, the design and subsequent application of the technology is, as it should be, dictated by the specific scientific questions being addressed. Thus, there can be no single design suitable for all scientific purposes. Another report of the Working Group will describe, in some detail, various types of facilities that have been constructed and used to address a variety of scientific questions.

The concept and approaches in using mesocosms are important in that many rates processes - biological and/or chemical interactions, for example -

cannot be quantified in any other way. As a general guide, it is my opinion that questions posed must be carefully identified and that the use of mesocosms be restricted to addressing the types of questions which cannot be answered from studies in the natural environment or in smaller, much cheaper, enclosures (microcosms). Examples of such questions are posed in several papers included in this report.

The purpose of using mesocosms is to eliminate some natural oceanographic variables (e.g. dilution by advection). It is important to understand, at the same time, that several other important natural processes are also eliminated within enclosures. For this reason, it is not reasonable to expect that mesocosms can, or should, exactly mimic the natural environment. It would be surprising if they did. The most important variables eliminated in any artificial enclosure, in terms of ecosystem balances, may be blocking the natural recruitment of organisms, natural inputs of nutrients, vertical mixing, etc. In experiments with pollutants, for example, the immediate effects of toxic materials on the contained organisms can be determined but, lacking recruitment, much less can be learned about how long it may take for affected populations to recover from stress. Also, "overexposure" of the organisms to the pollutant may result in a situation where concentrations are not diluted by advective and mixing processes. The noted effects, therefore, may be maximal rather than typical of the open ocean environment.

There have been frequent criticisms of mesocosm research, usually on the grounds that costs are too high relative to the scientific return. To pose and accept such criticism it would be necessary, at the same time, to accept the premise that the questions addressed using mesocosm technology have not been worth asking or answering. Examples of the application of such technology, to obtain answers to questions that could not have been obtained in any other way, are abundant throughout this report. Scientists using the technology have and will continue to provide important inputs to our understanding of how the oceans function.

David Menzel
Skidaway Institute of Oceanography

CONTRIBUTORS AND MEMBERS OF S.C.O.R. WORKING GROUP 85

Li Guanguo (Chairman)
Ocean University of Qingdao, Qingdao, Shandong Province, The People's Republic of China

Timothy R. Parsons (Vice-Chairman)
Department of Oceanography, University of British Columbia, Vancouver, B.C. V6T 1W5, Canada

Torgeir Bakke
Norwegian Institute for Water Research, N-0808 Oslo 8, Norway

Uwe Brockmann
Institut für Biochemie und Lebensmittelchemie, Universität Hamburg, 2000 Hamburg 13, Federal Republic of Germany

John C. Gamble
DAFS Marine Laboratory, Victoria Rd., Aberdeen AB98DB, Scotland, U.K.

Pierre Lasserre
Station Biologique de Roscoff, Place Georges Teissier, 29680 Roscoff, France

Victor Øiestad
Institute of Marine Research, N-5024 Bergen, Norway

Michael E. Q. Pilson
Graduate School of Oceanography, University of Rhode Island, Narragansett, R.I. 02882-1197, U.S.A.

Sigurd Schulz
Institut für Meereskunde, DDR-2530 Rostock-Warnemunde, German Democratic Republic

Masayuki Takahashi
Botany Department, University of Tokyo, Hongo, Tokyo 113, Japan

Peter A. W. J. de Wilde
Netherlands Institute for Oceanic Sciences, 1790 Ab den Burg, Texel, The Netherlands

ACKNOWLEDGMENTS

The contributing authors of this book wish to acknowledge the Scientific Committee on Oceanic Research for establishment and support of Working Group 85, and for financial assistance in the preparation of this book. They also wish to thank the Department of Oceanography, University of British Columbia, for providing financial assistance and space. Financial support was also provided by NSERC Grant No. 0GP0006689 to T. R. Parsons. The editor is grateful for the willing and capable assistance provided by B. Rokeby.

TABLE OF CONTENTS

1	Introduction	1
2	Different Types of Ecosystem Experiments *Li Guanguo*	7
3	Marine Microcosms: Small-Scale Controlled Ecosystems *Pierre Lasserre*	20
4	Pelagic Mesocosms: I. Food Chain Analysis *Masayuki Takahashi*	61
5	Pelagic Mesocosms: II. Process Studies *U. Brockmann*	81
6	Benthic Mesocosms: I. Basic Research in Soft-bottom Benthic Mesocosms *P. A. W. J. de Wilde*	109
7	Benthic Mesocosms: II. Basic Research in Hard-bottom Benthic Mesocosms *Torgeir Bakke*	122
8	Specific Application of Meso- and Macrocosms for Solving Problems in Fisheries Research *Victor Øiestad*	136
9	Application of Mesocosms for Solving Problems in Pollution Research *Michael E. Q. Pilson*	155

10 Baltic Sea Eutrophication: A Case Study using Experimental Ecosystems 169
Sigurd Schulz

11 Mesocosms: Statistical and Experimental Design Considerations 188
John C. Gamble

12 The Use of Mathematical Models in Conjunction with Mesocosm Ecosystem Research 197
T. R. Parsons

Subject Index 211

1. INTRODUCTION

The Scientific Committee on Oceanic Research established Working Group 85 (SCOR WG.85) to review and assess the applications of studies conducted in experimental ecosystems to basic and applied oceanographic research. The three specific terms of reference given to SCOR WG.85 were as follows:

(1) to examine previous studies involving experimental ecosystems, and critically evaluate the results and the application of such techniques to estuarine, coastal, and open sea problems;

(2) to make recommendations for complete-systems approaches (mesocosms, field, laboratory, and simulation modelling) to current problems in biological oceanography; and

(3) to specify design criteria pertinent to studies in the range of estuarine, coastal, and open sea conditions.

At the first meeting held at the University of Hamburg in June, 1988, the Working Group specified the definition of the term "experimental ecosystem". Because the experimental designs should encompass a complete-system approach, it was agreed that an experimental ecosystem must meet certain criteria. It should be:

> physically confined; self-maintaining; multitrophic; have a duration time exceeding the generation time of the penultimate trophic level present; and of a size sufficient to enable pertinent sampling and measurements to be made without seriously influencing the structure and dynamics of the system.

For practical purposes, experimental ecosystems were also classified according to size as being:

> microcosms (<1 m^3);
> mesocosms (between 1 m^3 and 10^3 m^3); or
> macrocosms ($>10^3$ m^3).

On a two-dimensional scale, 0.25 m^2 surface area was set as a convenient way of delimiting the difference between benthic microcosms and benthic mesocosms. Characteristically, microcosms tend to be laboratory-bench systems, mesocosms are constructed field enclosures, and macrocosms are dammed natural basins.

This volume marks the completion of the first term of reference. It presents a compilation of previously published work resulting from experiments with enclosed ecosystems of various sizes and designs. Both pelagic and benthic ecosystems are represented, as well as both basic and applied applications.

The second term of reference has been given preliminary discussion by Working Group 85. As the aim of that term of reference is to provide guidance for the future application of experimental ecosystem techniques, consideration has been given to current problems in biological oceanography that may be particularly suited to study by these means. Whereas the purpose of the Working Group was to consider only applications in biological oceanographic research, the group is aware of the value of such large experimental systems in other disciplines. For example, it is obvious that research on subjects such as marine geochemistry is exceedingly relevant to biological oceanography, and the Working Group thus hopes that its work and findings will be of some use to researchers in chemical and physical oceanography.

To date, most research using experimental ecosystems has been carried out in the developed countries, but the Working Group feels that there is a need to apply the methodology to appropriate problems in biological oceanography in many other regions throughout the world.

Current problems in biological oceanography that could be approached through experimental ecosystem research fit into three broad categories:

(1) general biological oceanography, particularly the need to understand processes and to measure fluxes;

(2) aspects of exploitation and management of resources; and

(3) fates and effects of pollutants.

In considering the current problems in biological oceanography, the broad categories need to be focused down to pertinent questions of the moment, such as the relationship between biological and physical processes in the sea. Other pertinent research issues are outlined below. There is also a great need to persuade biochemists and physiologists to work in experimental ecosystems, particularly on the more detailed ecosystem interrelationships that can only be followed in an enclosed situation.

Specifically, experimental ecosystems must now be accepted as a valid tool in oceanographic research, and as a tool which is particularly useful for certain aspects such as the estimation of benthic secondary production, the examination and validation of current descriptive ecosystem models, or the detailing of processes determining succession including, for example, the circumstances leading to algal blooms.

However, scientists working with experimental ecosystems must be able to define the insufficiencies and shortcomings of the method. If necessary, this must be done by parallel studies in the field. In fact, the use of experimental ecosystems must be seen as an adjunct to surveys and studies in nature. Figure 1 illustrates the general connection between experimental ecosystem research and studies in the natural environment and in the laboratory, using models where appropriate. This figure endorses both the holistic and reductionist approach to ocean research. In the context of using interrelated approaches to research problems, experimental ecosystems are particularly useful for:

(a) testing hypotheses (i.e. true manipulative experiments);
(b) carrying out exploratory investigations that are carefully planned; or
(c) creating specifically designed experiments involving the use of hazardous substances.

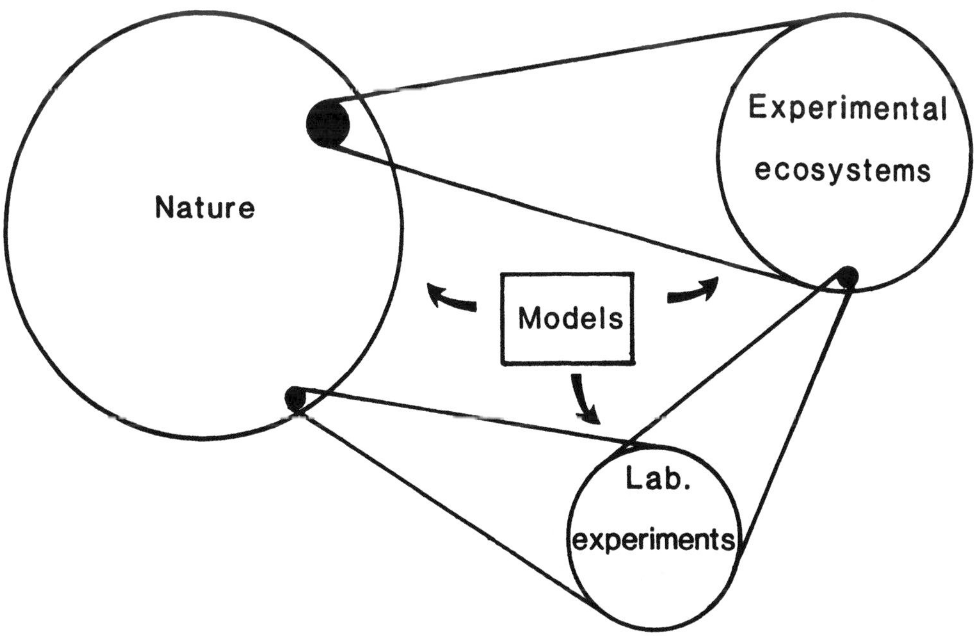

Figure 1. Position of experimental ecosystems in the framework of oceanographic research methodology.

The following constitute a series of research problems for which experimental ecosystems are thought to be appropriate:

I. General Biological Oceanography

I.1. The development of populations and biological systems, such as succession or recruitment

I.2. The transfer and fluxes of organic carbon, nitrogen, and phosphorus as, for instance, in uptake and mineralization

I.3. The role and biological significance of living and nonliving particulate material and of dissolved organic matter

I.4. The measurement and elucidation of ecosystem feedback and control mechanisms; for instance, exchange kinetics and physiological succession, phasing, and seasonal timing. These aspects lead directly into modelling and to the validation and testing of ecosystem models.

I.5. Boundaries and fluxes - experimental ecosystems have great potential for examining these most important aspects of biological oceanography:

-the air-water interface; photochemical processes; surface-film biology and chemistry;

-water column vertical structure; simulation of the creation and destruction of stratification;

-the benthic boundary; pelago-benthic coupling; and bioturbation and physical disturbance.

II. Exploitation and Management of Resources

With regard to fisheries questions, experimental ecosystems have been demonstrated to be particularly useful for investigations on the biology of fish larvae with respect to feeding rates, growth rates, and interaction between species, and to the measurement of mortality due to predation, toxic algae, and other agencies. Such data have important implications for the understanding of recruitment mechanisms.

Very little is understood about the biology and significance of juvenile fish (i.e. postlarval stages). It should be possible to devise very large macrocosm-scale systems for such investigations. It should also be possible to apply aquacultural techniques, both in the utilization of currently available large-scale systems and in the adoption of tried and tested engineering structures. There should also be much scope in investigating the use of man-modified lagoon environments for recruitment studies. This might be particularly relevant in tropical areas and raises the general question of inter-regional comparisons of ecosystems.

III. Pollution

Experimental ecosystems must now be regarded as being essential tools in studying marine pollution to bridge the gap between single-species toxicity tests and the natural environment. Emphasis should be directed toward:

(a) eutrophication, particularly with respect to algal blooms and truncated or enhanced succession;
(b) the fate and effects of complex pollutants such as harbor sludge, industrial effluents, river discharges, and oil derivatives;
(c) the temporal transformation of pollutants;
(d) the sequestering of pollutants into different phases including bioavailability, food-chain transfer, and biomagnification;
(e) the development of realistic simulated scenarios for specific purposes; and
(f) the examination of long-term chronic effects of pollutants on ecosystems.

Preliminary discussion on the third term of reference regarding specifications of design criteria has led to acknowledgment that experimental ecosystems must be designed to meet the particular demands of the envisaged needs. The degree of complexity proposed must match the scientific objectives of the exercise. However, all potential users should first ask, "Is an experimental ecosystem really necessary?"

Designs can be tailored to the needs and resources of local scientific communities, especially if such systems are to be used by developing countries. Proposed designs should include an indication of the advantages and disadvantages of the recommended systems, particularly the scientific and operational limitations. However, there must be awareness that the level of construction can be simple, but the analysis of results obtained will be complex and sophisticated and is highly dependent on skilled manpower.

Designs should include both structural aspects and a conceptual framework for the systems. Ideally, experimental ecosystems should be part of a suite of techniques applied during an investigation. In particular, they should complement parallel field measurements. Similarly, when specific hypotheses are being tested, much attention should be applied to careful statistical and experimental designs.

Future developments are likely to include new technology, and it must be expected that a new generation of experimental ecosystems could be more sophisticated, more automated, and more expensive. Such developments might be pursued through industrial involvement.

Two specific designs have been considered for which future needs are envisaged:

(a) Free-floating, open-water systems where the main problem is longevity and survival: For practical purposes, such systems should be subsurface to avoid wave action. This will create great sampling problems and, for the present and unless advanced technology is to be applied, such systems can only be used for short periods in calm water. It must be emphasized, however, that there are many uses for such limited systems.

(b) Sublittoral benthic systems: There are presently great problems in the simulation and translocation of a representative, deep-water, benthic community. From what depths can properly functioning samples be taken, and how realistic are ambient pressure simulations of such systems?

Working Group 85 is presently preparing a manual in which all currently known experimental ecosystems will be illustrated, and design features and uses described. The relative merits and drawbacks of each system will be emphasized, and it is hoped that the manual will serve as the guideline for future users of experimental ecosystems.

2. DIFFERENT TYPES OF ECOSYSTEM EXPERIMENTS

Li Guanguo

This paper is not intended as a comprehensive classification of the various types of ecosystem experiments in existence or of those suggested by some scientists. The different systems have been reviewed in papers by Kinne (1976), Menzel and Steele (1978), Pilson and Nixon (1980), Giddings (1981), Grice and Reeve (1982a), and in relevant chapters in a book edited by White (1984). In this paper, I will examine some of the problems and concepts in this broad field of work. Controversy is expected, but I hope that the ensuing discussion will be of value to the fulfillment of the Working Group tasks as they were defined by the terms of reference.

Ecosystem experiments have been initiated and developed to meet the needs of three fields of studies:

i) exploitation and management of marine resources;
ii) marine pollution and environmental protection; and
iii) biological oceanography.

The differences in objectives and designs of experiments in these three fields are very obvious, and there will be other papers to review them in this volume. The development of ecosystem experiments has been greatly influenced by the rapid development in instrumentation or technology and by new theories in closely related sciences, especially systems theory and systems analysis (including modelling and simulation). These facts have been pointed out first, because they should be kept in mind when reviewing or evaluating the various experiments. For ease of presentation, the following material has been drawn mainly from experiments on pelagic systems.

MESOCOSMS VERSUS MICROCOSMS

It is rather amazing to find that the word "mesocosm" was so readily accepted after its first appearance in the book "*Marine Mesocosms*" edited by Grice and Reeve (1982a). According to Grice and Reeve (1982b), the word "mesocosm" was proposed by Banse to cover the size range of experimental vehicles larger than benchtop containers but smaller than, and isolated from, any subunit of the natural environment. Santschi (1982) stated that "mesocosms lie between the complex and highly variable natural world and the tightly controlled but less natural laboratory experiment." These two definitions have brought four features of mesocosms to our attention: (i) their large size; (ii) the fact that they are less variable and (iii) less complex than the natural world; and (iv) that they are more natural and not so simple as microcosms. We might gather from papers produced by mesocosm workers some other features, such as the duration of the experiments, etc.

It is common knowledge that larger organisms require larger containers for their "normal" lives in captivity. Parsons (1982) has summarized relevant information for enclosure experiments in a simple diagram (Fig. 1). However, the size range of mesocosms has not been clearly defined. Banse (1982), in his historical review, set the lower limit at 1 m^3. Grice and Reeve (1982b) set a minimum volume of approximately 10 m^3 to exclude laboratory-scale experimental designs. It is true that some mesocosm experiments have used enclosures holding more than 1,000 m^3 of sea water, but there are also laboratory experiments using containers holding much more than 1 m^3 of sea water. So there is no absolute separation between mesocosms and microcosms in this respect.

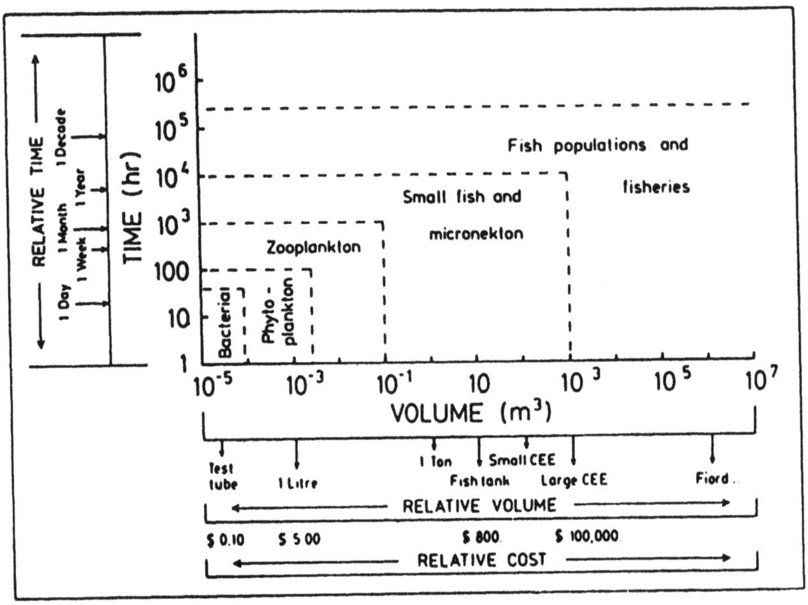

Figure 1. Approximate relationships between aquatic organism life cycles and the relative size and cost needed for their containment. CEE, Controlled Experimental Ecosystem. From Parsons (1982).

Kuiper et al. (1983) made a special study of the influences of bag dimensions on the development of enclosed plankton communities using bags containing 1.5 to 30 m^3 of water, and they concluded that optimum dimensions depend on the aim of the experiment, the number of trophic levels enclosed, the population densities at these levels, and on the species present at these levels. According to them, enclosures of 1 to 2 m^3 are sufficiently large for ecotoxicological experiments with marine phyto- and zooplankton communities in eutrophic waters; in more oligotrophic waters, enclosures with a volume of about 10 m^3 are preferred. There is no doubt that size of the enclosure is one of the important factors in the design of ecosystem experiments. However, there are other factors that are also important or even more critical.

The use of large enclosures in the sea has solved the problem of sampling and studying the same assemblage of natural populations for the whole time of the experiment. This problem has greatly troubled field workers because constant water movement and mixing are unique properties of the sea that mingle spatial variability and temporal changes, making the study of a particular natural assemblage of populations impossible. It is true that enclosing a natural assemblage of populations introduces unnatural conditions into an experiment, but this is inevitable. As Pilson and Nixon (1980) pointed out, "We cannot overemphasize that no microcosm can ever be an exact replica of nature for several reasons." (At that time, the word "microcosm" was used for both mesocosm and microcosm.) I have run across a paper with a comment that, with whatever improvement the mesocosm could have had, it would remain a "beaker to the bay." This fact should not make us waver in our evaluation of the introduction of large enclosures into ecosystem experimentation as a significant advance.

There has been a tendency to judge the complexity of an experiment by the number of trophic levels it contains. This practice or tendency has as its basis the concept of trophic-level formalism introduced by Lindeman (1942), which has had great influence for quite a long period in community and ecosystem studies, including ecosystem modelling work. Platt (1985) has cited Cousins (1980) and thinks that the introduction of this formalism might be considered in some respects to have been a retrograde step from the important earlier advances made by Elton (1927) and Hardy (1924). I myself have wondered for quite some time if the oversimplified trophic-level concept might be the stumbling block to marine ecosystem studies. I will come back to this topic later. Here I would like to point out that, with the rapid development of microcosm designs, complexity in the sense commonly used now may not be a suitable criterion for separating mesocosms from microcosms (cf. Taub 1984a, 1984b; Pritchard and Bourquin, 1984).

The possible duration of an experiment usually depends upon the size of the container and mode of operation. In laboratory work, a batch culture cannot be maintained very long, whereas a continuous culture can last much longer. According to Grice (1984), an enclosure for pollution research with more than two interacting trophic levels, each having more than one species, can be maintained for periods up to six or more weeks. Experiments with MERL tanks (flow-through type) can be run for more than a year. Banse (1982) believed that the MERL design offers a reasonable compromise in the conflict among work on plankton in beakers, large bags, or bays. I think that there is some truth in his comment.

In reviewing the literature available to me for the preparation of this paper, I found that many of the scientists engaged in microcosm or mesocosm work have the same view that interactive laboratory microcosms, mesocosms, and field studies multiply the strength of each approach and are necessary for ecosystem studies (e.g. Oviatt, 1984; Taub, 1984a). In fact, many of the mesocosm workers in the past have used laboratory experiments run simultaneously with their enclosure experiment to help in explaining some of the events (e.g. Grice et al., 1980; Harris et al., 1982).

FROM SINGLE-SPECIES CULTURE EXPERIMENTS TO "SYNTHETIC MICROCOSMS" OR "MODEL ECOSYSTEMS"

Single-species experiments (axenic or non-axenic, with food organisms or without food organisms in the case of zooplankters) are usually excluded from ecosystem experiments because they are used mainly for physiological or ecophysiological studies. However, their important contributions to ecosystem studies should not be overlooked (cf. Strickland, 1967). This type of experiment has been done successfully with plastic enclosures at sea (e.g. Brockmann et al., 1977; Eberlein and Brockmann, 1986; and many others); some experiments were also used to study chemical processes.

Mixed populations of different species of phytoplankton (one trophic level) have been used in many laboratory experiments to study such processes as competition, succession, etc. The first experiments with large plastic spheres were done with mixed populations of phytoplankton (McAllister et al., 1961; Antia et al., 1963).

Multispecies experiments with species from two or more trophic levels are very common in laboratory work. Some of them have been done with large enclosures (e.g. Mullin and Evans, 1974). Most of the experiments have been gnotobiotic.

Experiments with the highest complexities are those conducted in mesocosms and microcosms as discussed in the previous section. However, there is one type of experimental ecosystem that is very different from the others. It is what Nixon (1969) called a "synthetic microcosm", probably because it is established by synthesizing or assembling a model ecosystem from organisms (usually from convenient laboratory stocks of single-species cultures) that represent important roles in aquatic communities (Leffler, 1984; Taub, 1984b). Some of these experiments have been totally gnotobiotic (i.e. the total species assemblage including bacteria was defined). Strickland (1967) concluded that "probably the most valuable aspect of working with artificial marine ecosystems is that one can test, quantitatively, certain hypotheses made during field studies and, conversely, the behaviour of simplified systems leads to theories to be tested in the field." I think that this type of experiment has great potential to help us find some new approach to ecosystem experiments, which will be the topic of the following section.

VENTURE INTO A NEW POSSIBLE APPROACH

At this point, it might be necessary to review some of the important concepts and facts relevant to the study of marine ecosystems. The biosphere is characterized by the interrelationships of living things and their environments. Communities are interacting systems of organisms tied to their environments by the transfer of energy and matter. Such a coupling of living organisms and the nonliving matter with which they interact defines an ecosystem. An ecosystem may range in size from a small pond to a bay...to a shallow sea...to the entire biosphere. These ecosystems of different sizes and levels of complexity and relative independency constitute a hierarchical organization. This concept is consistent with the modern systems theory and the historical development of the biosphere. The marine ecosystem, which comprises the world's oceans with their shores and estuaries, is tremendous in size and complexity. To study it, we need to have some knowledge or hypothesis of its hierarchical organization so that we can choose a specific subsystem of lower level to begin our study more realistically. This practice has been followed by marine ecologists to some extent in the past as, for example, in studies of different coastal ecosystems (Mann, 1982). The question I should like to bring out in this section is whether we have done rightfully in this respect with our ecosystem experiments on pelagic ecosystems.

Nearly all the enclosure experiments done in studying a pelagic ecosystem conceived it as a system with its biotic components lumped into compartments by trophic levels. The inadequacy of lumping species of greatly different sizes into one compartment has been noted by many scientists (cf. Platt et al., 1981). Unfortunately, this practice has also been commonly adopted by field workers and modellers. The development in the study of particle-size distribution in sea water (cf. Sheldon and Parsons, 1967; Sheldon et al., 1972) opened a new field and a potential new approach for pelagic ecosystem studies. Platt (1985) called our attention to the important contributions made a long time ago by Elton (1927). As Platt summarized in his paper:

> "Appreciation of the value of thinking about the ecosystem in terms of the size-spectrum of the organisms that compose it can be traced to Elton (1927). He devoted an entire chapter of his textbook on ecology to 'the animal community' in which the conceptual thread was, explicitly, organism size. Elton's foresight was remarkable; in this one chapter he formulated all of the principles on which the modern theory of the pelagic ecosystem is based. He noted that animal communities can be conceived of as a series of interconnecting food chains forming a food web; he noted that smaller animals are much more abundant than larger ones, and characterised this as the 'pyramid of numbers'; he noted that the size of an organism's food bore relationship to its own body size, and concluded that the elemental food chains comprising the food web are arranged in order of increasing organism size, with the result that the general flow of material through the community is from small organisms to large ones; and finally he observed in a cryptic but pregnant final sentence, that the biomass of prey necessary to sustain a given predator was a function of both the food requirements of the predator and generation time of the prey."

From this quotation and some other ideas discussed in Platt's paper, I have a bold notion that it might be worthwhile to free ourselves from the yoke of the oversimplified trophic-level formalism and venture into the question of whether the pelagic ecosystem is actually formed by interdependence and interaction of many subsystems (Jørgensen, 1983a; 1983b).

Sieburth et al. (1978), in a study of pelagic ecosystem structure, tried to regroup the heterotrophic marine plankton into different compartments on the bases of the level of organization (ultrastructure) and the mode of nutrition, and then equated these with plankton size fractions (Fig. 2). They removed the apochlorotic flagellates from the phytoplankton, and the ciliates and the amoeboid forms from the microzooplankton, and placed them together into a new grouping, the protozooplankton. The microphages of this group prey mainly on the small procaryotes, the bacterioplankton. Johannes (1965) showed in culture experiments that the remineralization rate per unit weight of protozoa, both flagellates and ciliates, is at least an order of magnitude higher than that of most metazoan plankton. Johnson et al. (1982) studied a photosynthetic cyanobacterium, *Synechococcus* spp., a picoplankton ranging in size from 0.6 to 1.0 μm in diameter. They found from culture experiments that a species of *Actinomonas*, a helioflagellate measuring 3 x 5 μm , could consume an oceanic isolate of *Synechococcus* and be maintained on phototrophic bacteria for over one year. In similar experiments, the ciliate

Uronema, also isolated from Narragansett Bay, was fed and maintained on a growing oceanic isolate of *Synechococcus*. Electron microscope observations on the gut and fecal pellets of calanoid copepods from areas with maximal cyanobacteria populations showed that these cells were ingested in good numbers by the copepods. Whereas *Calanus finmarchicus* can feed on particles in the 5 to 10 μm range, and whereas it and other species of *Calanus* primarily use particles larger than 20 μm, the presence of numerous cyanobacteria and chlorophytes less than 3 μm in size in the fecal pellets of *C. finmarchicus* and other calanoids strongly suggests that they are ingested nonselectively. These cells are ultrastructurally unaltered, both in the gut and in voided fecal pellets. The authors presented a hypothetical model (Fig. 3) in their paper.

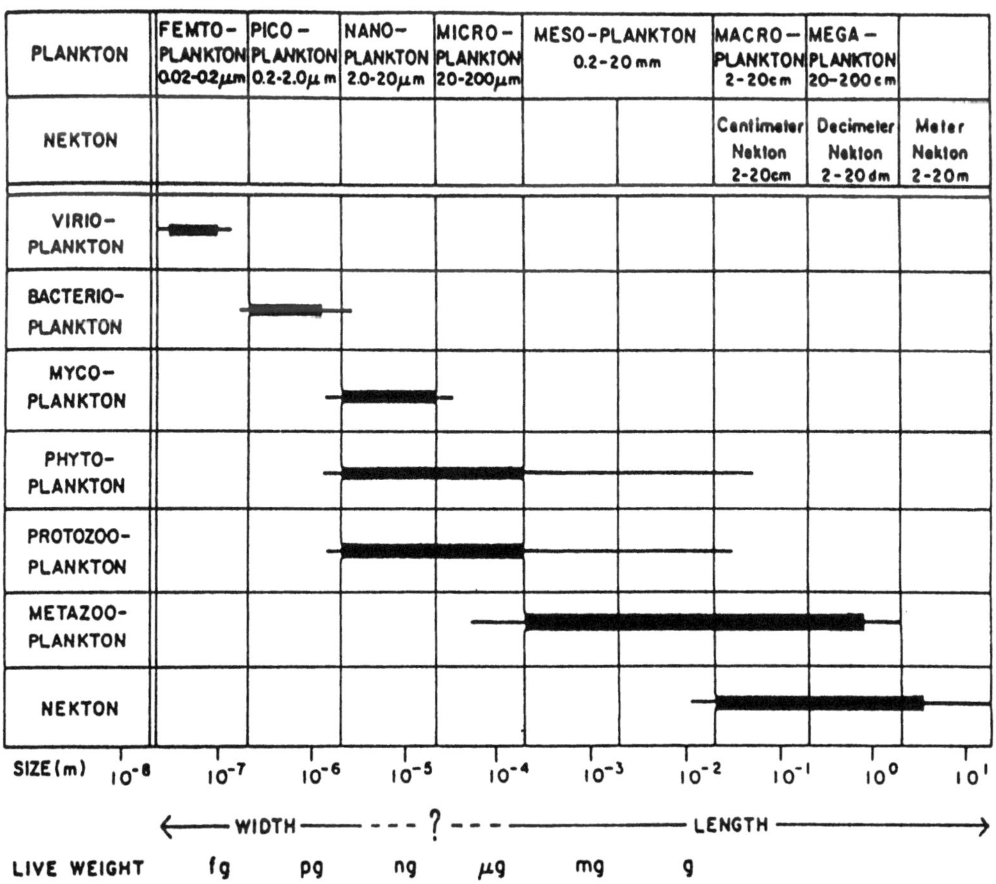

Figure 2. Distribution of different taxonomic-trophic compartments of plankton in a spectrum of size fractions, with a comparison of the size range of nekton. From Sieburth et al. (1978).

The model in Fig. 3 was made to show the role of chroococcoid cyanobacteria. It could be used as a part of a possibly existing subsystem with component organisms of the size ranges from picoplankton to microplankton (cf. Bratbak, 1987). We need to add into it other nanoplankton that can feed on bacterioplankton, and also microplankton that take nanozooplankton as their food. The boundaries of this subsystem can probably be set up by the trophic relationships within the assemblage, the high reproductive and low dispersal potentials of the biotic components, and the weak interaction with other subsystems having biotic components of larger size ranges. This microbial subsystem might be considered an inherent property of the sea water. The theories on the origin of life and the historical development of the biosphere probably could throw some light on this venture too.

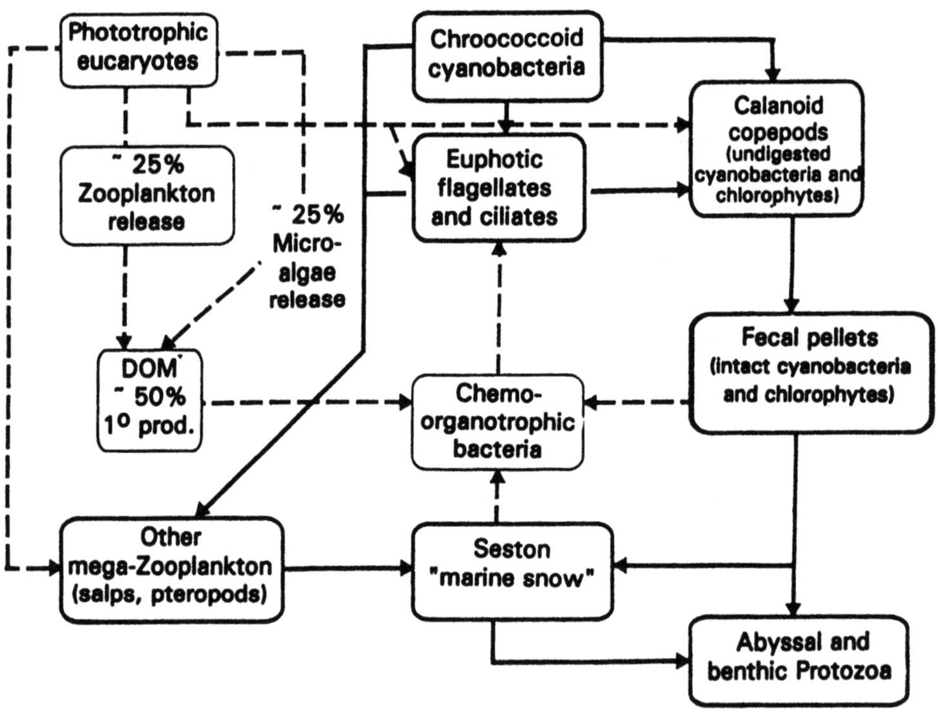

Figure 3. A hypothetical model of the role of chroococcoid cyanobacteria (boldface, solid lines) in the direct nutrition of euphotic-zone protozoa through fecal pellet transport, and how this route interfaces with the bacterioplankton and protozooplankton links for the recovery of lost productivity to the classic food chain (light face, dotted). From Johnson et al. (1982).

There is a great possibility of separating the other planktonic organisms into different subsystems too. The three types of communities (oceanic, continental shelf, and upwelled) with nanoplankton, microphytoplankton, and macrophytoplankton at the bases of the food chains are quite well accepted (e.g. Parsons et al., 1984). These three types of primary producers were found existing together in the enclosures used for the Foodweb I experiment, which was designed to test the hypothesis that the structure of the phytoplankton community determines the type and number of links in the food chain, and thus controls the transfer efficiency from primary production to upper levels such as fish (cf. Grice et al., 1980; Harris et al., 1982). This suggests that there were at least two possible subsystems in the water used for the experiment. It will need a lot of work to make out a tentative scheme. I believe that such a venture is hopeful. In order for different subsystems co-evolved over a long time to co-exist in the same water column, there must be some adaptive differentiation in food habits, feeding mechanisms, body size, behavior patterns, and so forth. There has been quite a lot of work done in the past on food relationships. What needs to be done now is to keep in mind the goal of finding possible subsystems and design experiments to test them. In doing so, we need field observations, experiments, and modelling (and computer simulation experiments) that are designed and carried out together. Maybe the biological oceanographers and ecologists should take the first step.

As to the nekton, the fishes, I have an idea that they might form a subsystem (or systems) by themselves, with interactions such as age-specific predator-prey relationships, and stock size and recruitment relationships, etc. This problem probably has to be solved by field observation and modelling (cf. Beyer and Sparre, 1983). A nekton subsystem (or systems) can be coupled with the other subsystems that have interdependency and interaction with it (or them).

If this new approach could be proved to be feasible or even closer to reality, it should bring us some breakthroughs in many aspects of our ecosystem studies. What an enclosure captures might be just a few subsystems, with some larger carnivores isolated from another subsystem (or systems); thus some of the events in the enclosure experiment might have other explanations. The flow chart for a pelagic ecosystem could have subsystems as compartments instead of the ones derived from trophic levels. We could have subsystems at a lower level of the hierarchical organization of a large complex ecosystem that are more specific and can be more easily comprehended to begin our studies. With the recent development of new instruments and technology, such as flow cytometry and continuous recording of many important environmental factors, and of ecosystem theories, it should be possible for us to begin a search and make some trials.

LITERATURE CITED

Antia, N. J., C. D. McAllister, T. R. Parsons, K. Stephens, and J. D. H. Strickland. 1963. Further measurements of primary production using a large-volume plastic sphere. *Limnol. Oceanogr.* **8**: 166-183.

Banse, K. 1982. Experimental marine ecosystem enclosures in a historical perspective. Pp. 11-24. **In**: G. D. Grice and M. R. Reeve [eds.], *Marine Mesocosms. Biological and Chemical Research in Experimental Ecosystems*. New York: Springer-Verlag.

Beyer, J. and Per Sparre. 1983. Modelling exploited marine fish stocks. Pp. 485-582 (esp. 541 ff). **In**: S. E. Jørgensen [ed.], *Application of Ecological Modelling in Environmental Management, Part A*. The Netherlands: Elsevier Scientific.

Bratbak, G. 1987. Carbon flow in an experimental microbial ecosystem. *Mar. Ecol. Prog. Ser.* **36**: 367-376.

Brockmann, U. H., K. Eberlein, G. Hentzschel, H. K. Schöne, D. Siebers, K. Wandschneider, and A. Weber. 1977. Parallel plastic tank experiments with cultures of marine diatoms. *Helgol. Wiss. Meeresunters.* **30**: 201-216.

Cousins, S. H. 1980. A trophic continuum derived from plant structure, animal size and a detritus cascade. *J. Theor. Biol.* **82**: 607-618.

Eberlein, K. and U. H. Brockmann. 1986. Development of particulate and dissolved carbohydrates in parallel enclosure experiments with monocultures of *Thalassiosira rotula*. *Mar. Ecol. Prog. Ser.* **32**: 133-138.

Elton, C. 1927. *Animal Ecology*. New York: Macmillan. 207 pp.

Giddings, J. M. 1981. Laboratory tests for chemical effects on aquatic population interactions and ecosystem properties. Pp. 23-91. **In**: A. S. Hammons [ed.], *Methods for Ecological Toxicology. A Critical Review of Laboratory Multispecies Tests.* Ann Arbor, Michigan: Ann Arbor Science Publ.

Grice, G. D. 1984. Use of enclosures in studying stress on plankton communities. Pp. 563-573. **In**: H. H. White [ed.], *Concepts in Marine Pollution Measurements*. University of Maryland: Maryland Sea Grant.

Grice, G. D., R. P. Harris, M. R. Reeve, J. F. Heinbokel, and C. O. Davis. 1980. Large-scale enclosed water column ecosystems. An overview of Food-web I, the final CEPEX experiment. *J. Mar. Biol. Ass. U.K.* **60**: 401-414.

Grice, G. D. and M. R. Reeve [eds.]. 1982a. *Marine Mesocosms. Biological and Chemical Research in Experimental Ecosystems.* New York: Springer-Verlag. 430 pp.

Grice, G. D. and M. R. Reeve. 1982b. Introduction and description of experimental ecosystems. Pp. 1-9. **In**: G. D. Grice and M. R. Reeve [eds.], *Marine Mesocosms. Biological and Chemical Research in Experimental Ecosystems.* New York: Springer-Verlag.

Hardy, A. C. 1924. The herring in relation to its animate environment. Part I. The food and feeding habits of the herring with special reference to the East Coast of England. *Fishery Invest., London,* Series II (3): 1-53.

Harris, R. P., M. R. Reeve, G. D. Grice, G. T. Evans, V. R. Gibson, J. R. Beers, and B. K. Sullivan. 1982. Trophic interaction and production processes in natural zooplankton communities in enclosed water columns. Pp. 353-387. **In**: G. D. Grice and M. R. Reeve [eds.], *Marine Mesocosms. Biological and Chemical Research in Experimental Ecosystems.* New York: Springer-Verlag.

Johannes, R. E. 1965. Influence of marine protozoa on nutrient regeneration. *Limnol. Oceanogr.* ***10***: 434-442.

Johnson, P. W., Huai-shu Xu, and J. McN. Sieburth. 1982. The utilization of chroococcoid cyanobacteria by marine protozooplankters but not by calanoid copepods. *Ann. Inst. oceanogr., Paris.* **58**: 297-308.

Jørgensen, S. E. 1983a. The modelling procedure. Pp. 5-15. **In**: S. E. Jørgensen [ed.], *Application of Ecological Modelling in Environmental Management, Part A.* The Netherlands: Elsevier Scientific.

Jørgensen, S. E. 1983b. Eutrophication models of lakes. Pp. 227-282. **In**: S. E. Jørgensen [ed.], *Application of Ecological Modelling in Environmental Management, Part A.* The Netherlands: Elsevier Scientific.

Kinne, O. 1976. Cultivation of marine organisms. Water-quality management and technology. Pp. 19-300. **In**: O. Kinne [ed.], *Marine Ecology, Vol. III, Cultivation*. London: J. Wiley.

Kuiper, J., U. H. Brockmann, H. van het Groenewoud, G. Hoornsman, and K. D. Hammer. 1983. Influences of bag dimensions on the development of enclosed plankton communities during POSER. *Mar. Ecol. Prog. Ser.* **14**: 9-17.

Leffler, J. W. 1984. The use of self-selected, generic aquatic microcosms for pollution effects assessment. Pp. 139-157. **In**: H. H. White [ed.], *Concepts in Marine Pollution Measurements.* University of Maryland: Maryland Sea Grant.

Lindeman, R. L. 1942. The trophic-dynamic aspect of ecology. *Ecology* **23**:399-418.

Mann, K. H. 1982. *Ecology of Coastal Waters. A Systems Approach.* Berkeley and Los Angeles, CA: University of California Press. 322 pp.

McAllister, C. D., T. R. Parsons, K. Stephens, and J. D. H. Strickland. 1961. Measurements of primary production in coastal sea water using a large-volume plastic sphere. *Limnol. Oceanogr.* **6**: 237-258.

Menzel, D. W. and J. H. Steele. 1978. The application of plastic enclosures to the study of pelagic marine biota. *Rapp. P.-V. Réun. Cons. Perm. Int. Explor. Mer* **137**: 7-12.

Mullin, M. M. and P. M. Evans. 1974. The use of a deep tank in plankton ecology. 2. Efficiency of a planktonic food chain. *Limnol. Oceanogr.* **19**: 902-911.

Nixon, S. W. 1969. A synthetic microcosm. *Limnol. Oceanogr.* **14**: 142-145.

Oviatt, C. A. 1984. Introduction: Ecology as an experimental science and management tool. Pp. 539-547. **In**: H. H. White [ed.], *Concepts in Marine Pollution Measurements.* University of Maryland: Maryland Sea Grant.

Parsons, T. R. 1982. The future of controlled ecosystem enclosure experiments. Pp. 411-418. **In**: G. D. Grice and M. R. Reeve [Eds.], *Marine Mesocosms. Biological and Chemical Research in Experimental Ecosystems.* New York: Springer-Verlag.

Parsons, T. R., M. Takahashi, and B. Hargrave. 1984. *Biological Oceanographic Processes.* Oxford, England: Pergamon Press. 330 pp.

Pilson, M. E. Q. and S. W. Nixon. 1980. Marine microcosms in ecological research. Pp. 724-741. **In**: J. P. Giesy, Jr. [ed.], *Microcosms in Ecological Research.* DOE Symposium Series 52, CONF-781101. Springfield, VA: National Technical Information Service.

Platt, T. 1985. Structure of the marine ecosystem: Its allometric basis. Pp. 55-64. **In**: R. E. Ulanowicz and T. Platt [eds.], *Ecosystem Theory for Biological Oceanography.* Can. Bull. Fish Aquat. Sci. 213.

Platt, T., K. H. Mann, and R. E. Ulanowicz. 1981. *Mathematical Models in Biological Oceanography.* Paris: The Unesco Press. 156 pp.

Pritchard, P. H. and A. W. Bourquin. 1984. A perspective on the role of microcosms in environmental fate and effects assessments. Pp. 117-138. **In**: H. H. White [ed.], *Concepts in Marine Pollution Measurements.* University of Maryland: Maryland Sea Grant.

Santschi, P. H. 1982. Application of enclosures to the study of ocean chemistry. Pp. 63-80. **In**: G. D. Grice and M. R. Reeve [Eds.], *Marine Mesocosms. Biological and Chemical Research in Experimental Ecosystems.* New York: Springer-Verlag.

Sheldon, R. W. and T. R. Parsons. 1967. A continuous size spectrum for particulate matter in the sea. *J. Fish. Res. Bd. Canada* **24**: 909-915.

Sheldon, R. W., A. Prakash, and W. H. Sutcliffe, Jr. 1972. The size distribution of particles in the ocean. *Limnol. Oceanogr.* **17**: 323-340.

Sieburth, J. McN., V. Smetacek, and J. Lenz. 1978. Pelagic ecosystem strucure: Heterotrophic compartments of the plankton and their relationship to plankton size fractions. *Limnol. Oceanogr.* **23**: 1256-1263.

Strickland, J. D. H. 1967. Between beakers and bays. *New Scientist* **33**: 276-278.

Taub, F. B. 1984a. Introduction. Pp. 113-116. **In**: H. H. White [ed.], *Concepts in Marine Pollution Measurements*. University of Maryland: Maryland Sea Grant.

Taub, F. B. 1984b. Measurement of pollution in standardized aquatic microcosms. Pp. 161-192. **In**: H. H. White [ed.], *Concepts in Marine Pollution Measurements*. University of Maryland: Maryland Sea Grant.

White, H. H. [ed.]. 1984. *Concepts in Marine Pollution Measurements*. University of Maryland: Maryland Sea Grant. 743 pp.

3. MARINE MICROCOSMS: SMALL-SCALE CONTROLLED ECOSYSTEMS

Pierre Lasserre

INTRODUCTION

Since pioneer works in the early 1960s (Odum et al., 1963; Beyers, 1963), laboratory-scale experimental designs are more and more considered as fundamental tools in the development of freshwater and marine ecological research. The word "microcosm" was widely used to designate laboratory-scale aquaria as well as large-scale enclosures (Giesy, 1980) until the term "mesocosm" was coined to designate large-scale enclosures outside the laboratory of volume exceeding 1 m^3 (Grice and Reeve, 1982). Therefore, the word "microcosm" is used today to designate a contained and often simplified marine community of volume not exceeding 1 m^3, in which controlled experiments can be performed. Such small systems are convenient to compare single species and multispecies properties as to metabolic patterns and physiological characteristics, without the difficulties of replication and sampling. Both micro- and mesocosm approaches have their advantages and difficulties; they should be considered as fully complementary. The ideal microcosm experiment is one in which the investigator manipulates one particular factor, whereas all others are allowed to vary naturally. The factor of interest is controlled, if necessary at a series of different levels. This is the reverse of the classical laboratory "test tube" experiment in which all factors but one are kept rigidly controlled.

The limit of 1 m^3 which differentiates microcosms from mesocosms is rather arbitrary, but this size limit represents levels of control ranging from systems that are completely controlled to systems subject to some stochastic fluctuations. There are some variables that cannot be adequately controlled in microcosms to make them analogous to natural systems. Nevertheless, an attractive attribute of microcosms is their replicability, which is amenable to statistical testing of the effects of manipulation.

Using a microcosm in studies of natural or perturbed ecosystems requires that the investigator be thoroughly familiar with the basic character of the ecosystem being studied. More than for a mesocosm, components chosen for inclusion in the microcosm are those recognizable by the investigator and necessary to address the objectives of the study.

Useful ecological experiments cannot be done without a great deal of preliminary work. As for mesocosms, the ecologist looks for patterns of correlation between the distribution and abundance of the different organisms, or between the organisms and their abiotic environment. By asking questions about the causes of the patterns, answers suggest themselves which can be formulated as testable hypotheses. Experiments can then be designed to test these hypotheses.

A deeper understanding of microcosm function depends on good data being obtained on the fluxes of organic compounds (particulate and dissolved forms), and on the rates of feeding, assimilation, respiration, excretion, and other metabolic processes. Very significant progress has been made in recent years in developing appropriate techniques for measuring physiological rates and ecological fluxes in marine systems. Techniques of interest for microcosm studies range from flow cytometry (Yentsch et al., 1983; Burkill, 1987) to oxygen microelectrodes and microcalorimetry, as well as many biogeochemical sensors and tracers (Pamatmat, 1965, 1984; Lasserre, 1976, 1984; Zeitzschel, 1981; Revsbech, 1983; Denman et al., 1985; Gustafsson, 1987; Blackburn and Sorensen, 1988), and many other techniques for studies of feeding and comparative biochemistry and physiology (see Hochachka and Somero, 1973; Platt, 1981; Platt et al., 1981; Vernberg and Vernberg, 1981).

The data-sets in microcosms are not merely static measures but represent dynamic ecosystem behavior patterns that are difficult to quantify by other means. Microcosms are amenable to hierarchical studies because their scale can be readily varied from a fraction of a liter to 1 m^3. Within short periods (months to hours), ecosystem-level responses can be observed through primary succession, steady state, perturbation periods, secondary succession, and possibly senescence. A proliferation of shapes and designs of microcosms has been designed ranging from vials of a few milliliters, chemostats, and gradostats to gnotobiotic systems, and aquaria, bags, tubes, containers made of glass, soft or rigid plastic, steel etc., all containing transplanted portions of natural ecosystems. It is not possible to give an exhaustive list of these microcosm set-ups here.

The purpose of the present contribution is not to review the large body of literature on microcosms (*stricto sensu*), but rather to examine the results of some studies to illustrate the directions of possible future studies and to show that microcosms can afford interpretations of general interest.

MULTISTAGE CONTINUOUS CULTURE SYSTEMS

From the Chemostat to the Gradostat

There are many examples in the literature in the last decade of the use of chemostats for growing microorganisms and phytoplankton from freshwater and marine environments. It must be stressed that a chemostat does not in any way mimic a natural environment. Developed by Monod (1950) and Novick and Sziland (1950), the theory of continuous culture is based on a physiological steady-state situation occurring when the concentrations of substrates within a culture are artificially maintained at constant values and populations grow at a fixed rate equal to the dilution rate. Once a steady state has been achieved, the nature of the culture is independent of time. Obviously, there is no parallel for such a system in nature.

The deliberate cultivation of mixed populations raised interesting questions concerning the outcome of competitive interactions for growth-limited substrates. Growth rate in a chemostat is regulated by the concentration of a limiting substance (Fig. 1). Under steady-state conditions, the dilution rate (D) can be shown to be equal to the growth rate (μ) when the cells divide at a constant rate. In the case of $D < \mu m$, where μm is the unlimited growth rate with the particular substrate used, the increase of the population is:

$$dx/dt = x\,[\mu(x) - D]$$

where x is the population density and $\mu(x)$ is the density-dependent growth rate. As x increases, $\mu(x)$ will decrease due to the competition for the substrate between the bacteria or phytoplankton cells.

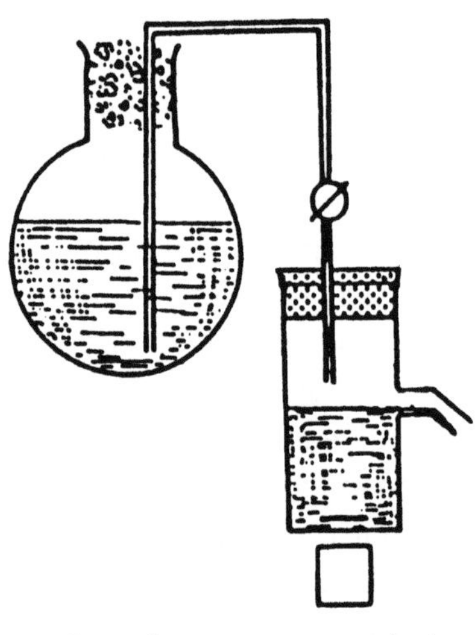

Figure 1. Schematic view of a chemostat consisting of a culture vessel with an overflow, and with the addition of a sterile substrate at a constant rate. From Christiansen and Fenchel (1977).

Mixed bacterial populations in a chemostat undergo selective processes. The species achieving the fastest growth rate will approach a steady-state population while the competing species are displaced. In a series of papers starting in 1953, Jannasch and his co-workers developed the concept of the chemostat as a laboratory model for natural aquatic systems, because the flow-through characteristics of the chemostat seemed to simulate certain aspects of the natural aquatic environment. Jannasch showed how bacterial competition can be simulated in the chemostat and explained the survival of "slow growers" by their greater efficiency of function at low nutrient concentrations.

The deliberate cultivation of mixed populations raises interesting questions concerning the competitive interactions of microbial communities. In chemostats, growth rate is regulated by the concentration of a limiting substrate; typical cases of Monod kinetics are illustrated in Fig. 2. In case (1), there is no possibility that organisms "B" can survive because, at all substrate

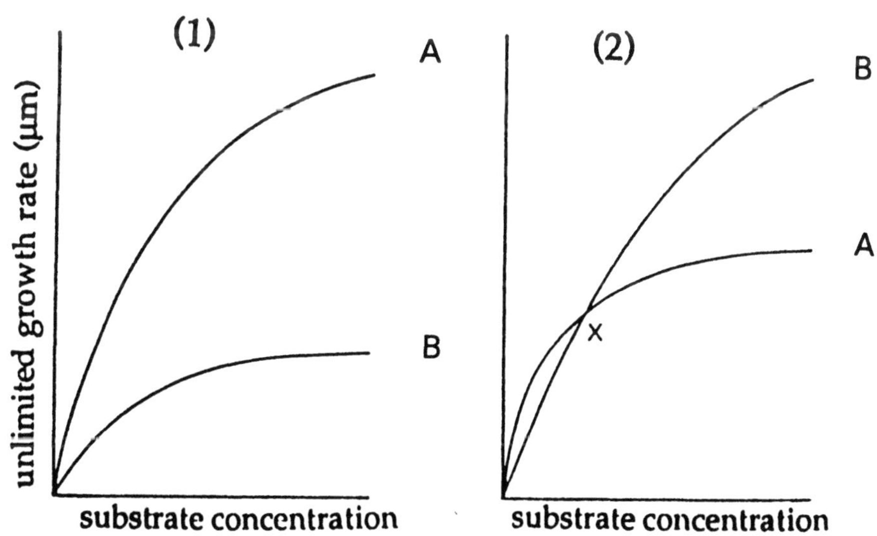

Figure 2. Growth rate as a function of substrate concentration in a chemostat for species A and species B, showing different Monod growth kinetics. (1) Population A outgrows population B at all substrate concentrations. (2) The outcome of competition between species A and B depends on the substrate concentration; species A will win at the low substrate concentration, but species B will be favored at a high substrate concentration. From Wimpenny (1981).

concentrations, organisms "A" will grow faster. In case (2), the outcome will depend on the substrate concentration, and therefore on the dilution rate. Below the crossover point of growth curves (X), organisms "A" will outgrow "B", but above it, the converse is true. At the crossover point, both organisms will theoretically survive. These two types of organisms are actually found in nature. For example, Jannasch (1967) has shown that the maximum rate of the bacterium *Spirillum* represents approximately the critical dilution rate, above or below which *Pseudomonas* or *Spirillum* will compete successfully (Fig. 3). Another example has been described by Veldkamp and Jannasch (1972) for two forms of photosynthetic sulfur bacteria growing in sulfide-limited chemostats. The form which wins at all optimal and suboptimal concentrations becomes dominated by the form which has "traded" high value of growth rate in the presence of high sulfide concentrations (Fig. 4).

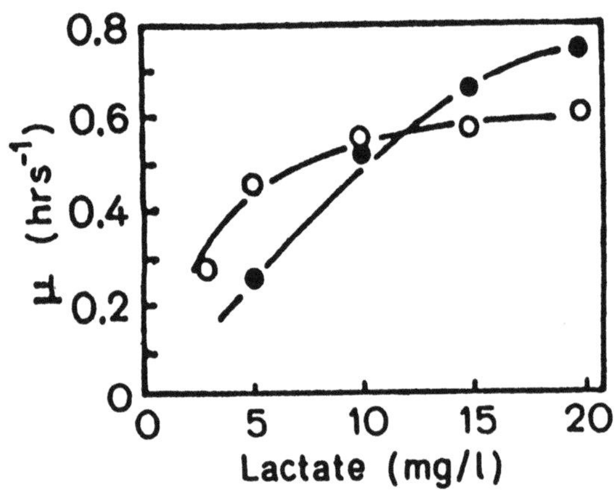

Figure 3. Growth rate in a lactate-limited culture of *Pseudomonas* (closed circles) and *Spirillum* (open circles). From Jannasch (1967).

Jannasch (1974) cautioned against the direct comparison of chemostat studies with the natural environment. He pointed out that "unambiguous terminology is needed for the proper use of simply continuous culture (batch) or the more sophisticated chemostat whenever applicable." The application of a mathematically-simplified model based on steady-state kinetics obtained in the chemostat may not lead to realistic analyses and predictions for the behavior of complex natural populations. Nevertheless, the great advantage of the chemostat is that the cell populations rapidly stabilize, giving a powerful tool for measuring various kinetics parameters of multispecies that co-exist as a stable association in a competitive situation. Responses of the species can be investigated at any growth rate, including extremely low rates which are found in some oceanic environments.

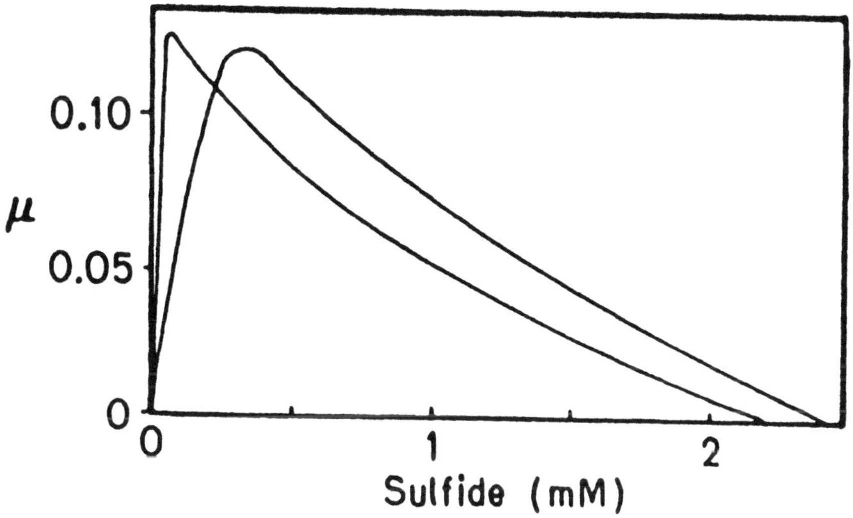

Figure 4. Growth rate of two photosynthetic sulfur bacteria as a function of sulfide concentration. After Veldkamp and Jannasch, 1972, in Christiansen and Fenchel (1977).

Although numerous papers based on field surveys have been written to identify the limiting nutrients in oceanic, neritic, estuarine, and lake studies, the data usually do not lend themselves to unequivocal interpretation. Since the work of Dugdale (1967) and Eppley and Coatworth (1968), chemostat studies, with continuous cultures maintained at steady state, have permitted dynamic aspects of nutrient uptake and assimilation to be examined with increasing precision (see McCarthy, 1981). Eppley et al. (1977) were the first to draw the analogy between the steady-state continuous culture and a segment of the open sea (in their case, the central gyre of the North Pacific Ocean). An *in situ*-type of chemostat (Fig. 5) has also been devised to study algal population changes (de Noyelles and O'Brien, 1974).

Beyond the problem of stating whether phytoplankton growth is nutrient limited or unlimited, the subsequent problem of estimating the importance of nutrient partitioning in species competition and succession has been underlined (Maestrini and Bonin, 1981). Only a few species have been involved in chemostat and batch experiments that demonstrate the algal capability to take up nutrient very rapidly. Many contributions (e.g. Conway and Harrison, 1977; Goldman et al., 1979; McCarthy and Goldman, 1979) deal mainly with experiments done with the diatoms *Thalassiosira pseudomonas* and *Skeletonema costatum*, both well known for their active metabolism which makes them convenient experimental material. Enhancement uptake rates at very low substrate concentrations, as reported by McCarthy and Goldman (1979), are suspected to have been overestimated. Moreover, little has been done to observe the concomitant behavior of two or more species sharing the same nutrient pool in the same experimental vessel.

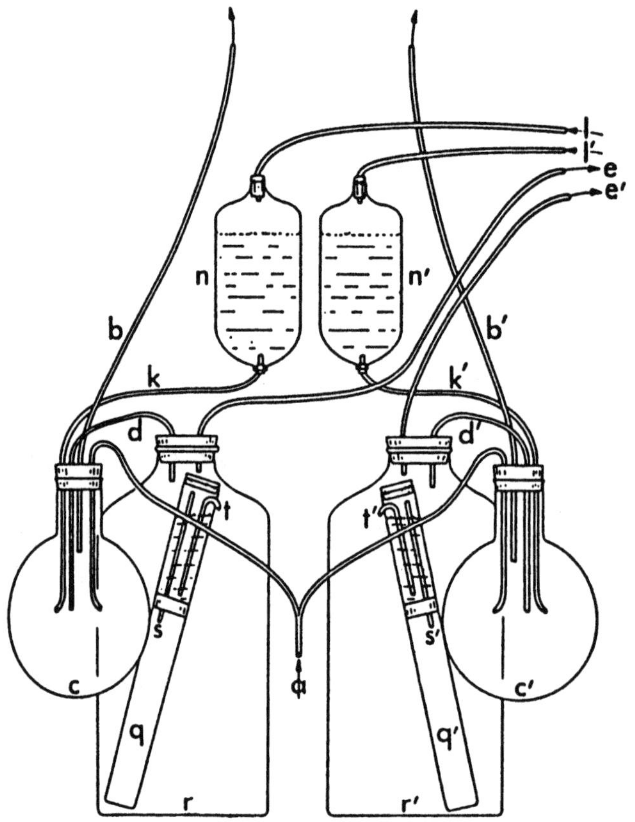

Figure 5. An *in situ* chemostat, a self-contained continuous culture unit showing duplicate systems for culturing (c and c'), collecting (r and r'), nutrient addition (n and n'), preserving (q and q'), and monitoring (e and e'). a, water inlet; b and b', air outlets; d and d', water connections between culture and collection chambers. From de Noyelles and O'Brien (1974).

Goldman (1984) stated that, in a chemostat and in the open ocean, both biomass and ambient nutrient remain invariant over time (Fig. 6). He pointed out that "nutrient input and biomass input are regulated by mechanical pumping in the continuous culture and, analogously, in a given portion of the open ocean these processes are controlled by the combination of grazing and nutrient regeneration."

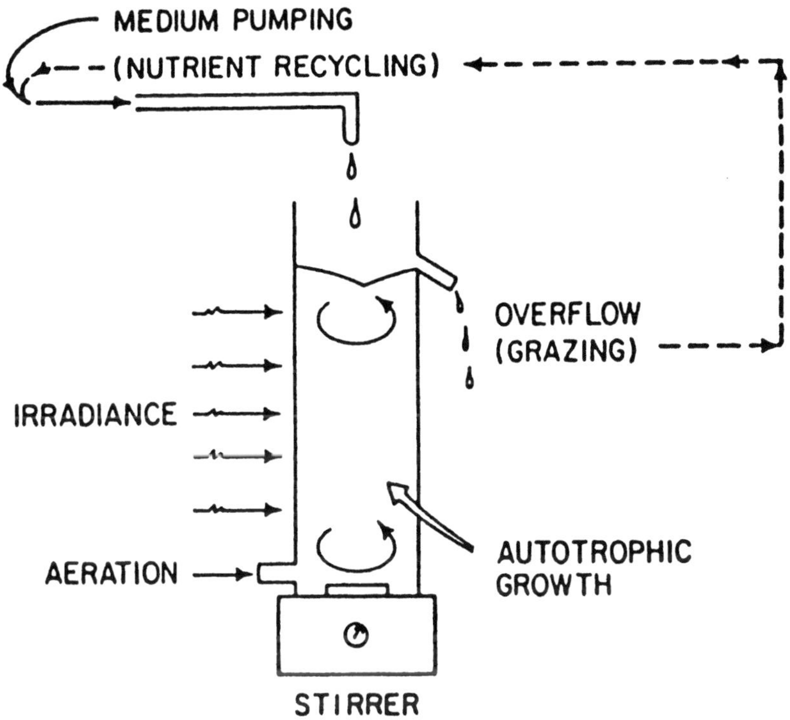

Figure 6. Schematic diagram of how a continuous culture can represent a segment of the open sea. From Goldman (1984).

To reconcile the seemingly incompatible concept that phytoplankton may be growing fast in oligotrophic waters with observations that indicate these waters are incapable of supporting such rapid growth, Goldman (1984) suggested that all small amorphous aggregates of organic matter in the water column may represent self-contained micro-habitats. Each aggregate is a floating "oasis" in the desert, serving as a life-support system for its resident population of autotrophs and heterotrophs (Fig. 7).

To better interpret field or mesocosm data on plankton ecology, we need increased microcosm studies with the organisms that are typical in natural assemblages. This will require more attention to the systematics of the phytoplankton - the nanoplankton in particular - and an increased effort to bring into culture more of the species that are commonly observed in nature, but rarely maintained in laboratories.

The development of the chemostat by microbiologists resulted in sophistication of continuous operations, such that steady-state relationships between microbial cultures and the environment could be attained. Because estuarine ecosystems can be considered to be "open" systems in which import and export are of paramount importance, continuous-culture methodology would appear to resemble estuarine dynamics more closely than would batch-type (aquarium, etc.) culture methods. This fact was recognized by Margalef (1963, 1967) when he mentioned the feasibility of utilizing chemostat methodology for laboratory estuarine simulations. He utilized such a device for studying aspects of plankton dynamics and succession in model estuarine systems. Although this system can provide useful information, it is unidirectional, and a better model of natural diffusion processes requires them to be bidirectional.

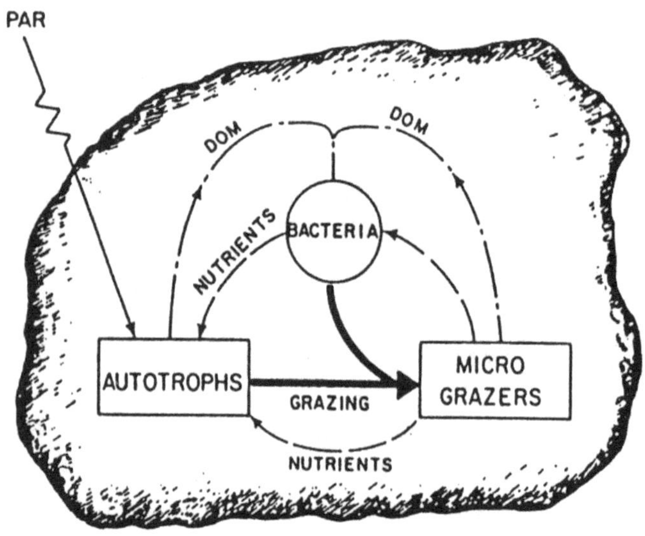

Figure 7. Conceptual diagram of a microbial loop colonizing a small amorphous aggregate of organic matter. Each aggregate serves as a life-support system for its resident populations of autotrophs and heterotrophs. From Goldman (1984).

A modified and elaborated version of the series of chemostats mentioned by Margalef (1967) was used by Cooper and Copeland (1973) to simulate the hydrological conditions of Trinity Bay, Texas, an estuarine ecosystem. The metabolic and structural responses of continuous-series microecosystems to changes in salinity gradient were investigated (Fig. 8). This model consisted of five linked and stirred vessels of 41 l each, with fresh water representing the river water input flowing in at one end. At the opposite end, salt water was introduced to represent the tidal inflow of sea water. This system established salinity gradients and resulted in the development of a classic estuarine planktonic flora. Retarding the freshwater input acted as an environmental stress on the first three cells of the microecosystems. Production and respiration were significantly lower, and zooplankton standing crop and species diversity decreased significantly. Addition of industrial effluent produced similar effects. Decreased freshwater input rate (primary stress) rendered the receiving communities more susceptible to the industrial effluent addition (secondary stress).

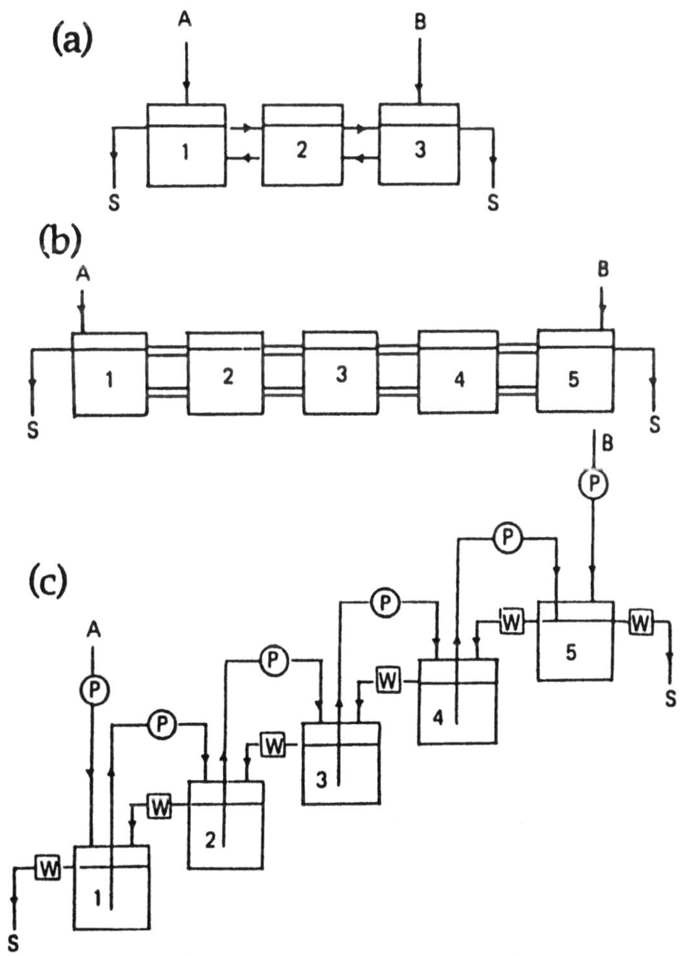

Figure 8. Schematic representation of different forms of the gradostat. (a) The principle of gradostat operation; (b) the estuarine model ("continuous-series microecosystems") of Cooper and Copeland (1973); (c) the gradostat of Lovitt and Wimpenny (1981): the five vessels are connected by weirs W and by tubing pumps P, and they are fed with medium from reservoirs A and B whilst S represents the sinks for each system. From Wimpenny (1981).

Lovitt and Wimpenny (1981) independently developed a similar system, termed a "gradostat". The system is based on a multistage chemostat principle, with the innovation that it uses bidirectional flow of media to generate opposite solute gradients. Flow in the gradostat, in contrast to the Cooper-Copeland model, is by peristaltic pumps in one direction and over weirs in the others (Wimpenny, 1981). The contents of the five vessels are transferred from right to left and from left to right simultaneously. Media enter the system from reservoirs located at both ends of the array. At the same time, culture leaves the system from the two end vessels. Opposing linear-stepped gradients for each solute will form under steady-state conditions (Fig. 8).

According to Wimpenny and Peters (1987), the gradostat is not a perfect model for investigating sediment systems because it has inputs at each end of the vessel array. They proposed a single-ended diffusion-coupled gradostat, as in Fig. 9 where the properties of a sediment bacterial community were investigated. The first vessel in which nutrients entered was aerated to simulate a superficial aerobic layer of sediment. Gradients of redox potential and of sulfate and sulfide concentration were all measured. Vessel 2 was characterized by the lowest redox potential (more metabolic activity). Sulfide concentration was highest in vessel 3, and low pH was observed in vessels 4 and 5. Sulfate-reducing bacteria were highest in vessels 3 and 4, and most aerobic and anaerobic heterotrophs were found in vessels 1 and 2 (Fig. 9). These results encourage the development of more research using this type of single-ended gradostat as a suitable model for natural spatially-heterogeneous communities, notably from marine ecosystems. A number of laboratory growth systems have been developed recently, notably by the research group headed by J. W. T. Wimpenny at Cardiff University College. These models incorporate spatial heterogeneity at one level or another. They range from open steady-state gradient systems, to gel-stabilized model systems, and to multidimensional gradient systems and microbial films. A version of the gradostat with more than five vessels is in preparation (Wimpenny and Peters, 1987).

Gel-stabilized Model Systems

In order to study the spatial distribution of physiologically distinct groups of bacteria, MacFarlane et al. (1984) used the gel-stabilized model system described by Wimpenny et al. (1981), but modified to simulate surface sediments of the Tay Estuary. The gel model was set up in a 1-l Quickfit reaction vessel. A gel medium was poured aseptically into the reaction vessel and allowed to set. Well-mixed sediment from the upper 5 cm was layered over the solid gel layer, and a semi-solid agar layer was aseptically poured over

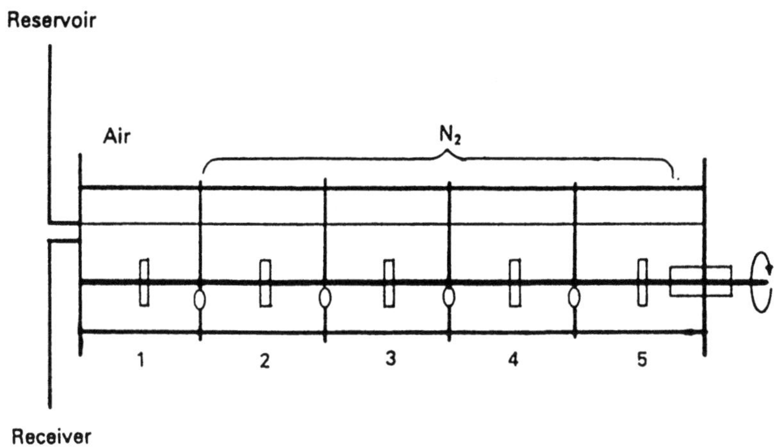

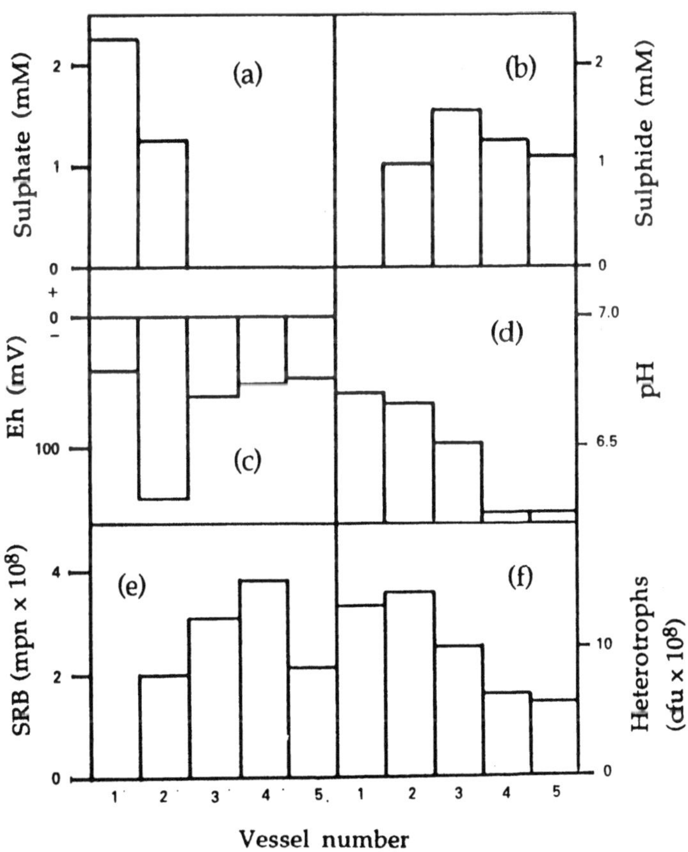

Figure 9. A new gradostat consisting of a single-ended diffusion-coupled system for studying the properties of a sediment-microbial community. A freshwater sediment community was introduced into the left-hand vessel and fed with a glucose-mineral salts medium; this vessel was aerated. The other four vessels were gassed with nitrogen. Parameters followed were (a) sulfate, (b) sulfide, (c) redox potential, (d) pH, (e) sulfate-reducing bacteria (SRB), and (f) total heterotrophs. From Wimpenny and Peters (1987).

this sediment. The gel was covered with aluminium foil and incubated at 20°C. The sole source of carbon in these models was that derived from the sediment layer. The semi-solid top layer of each gel was sampled at 0, 4, 8, 12, and 13 weeks, and the profiles of oxygen tension, Eh, pH, NO_2^-, NO_3^-, and NH_4^+ were determined together with the population densities of nitrifying, nitrate-respiring and sulfate-reducing bacteria.

The gradients of pH, Eh, and oxygen tension were similar to *in situ* measurements in Tay Estuary sediments. The spatial distribution of bacterial populations also showed a good correlation with field records in the surface sediments from Tay Estuary. This system is very different from the natural conditions in that the gel lacks bioturbation and the substantial amounts of particles that are present in natural conditions. Moreover, there is no gas or solute exchange. Nonetheless, it permits reliable analyses of microbial processes which are not easily obtained by other means. The gel-stabilized model shows, in effect, that NH_4^+ oxidation by autotrophic nitrifying bacteria plays an important role in the development of physico-chemical gradients within the gels, "indicating that similar processes may also operate *in situ* in these sediments" (MacFarlane et al., 1984).

Microbial Films

The development of biofilms in flow systems has been observed in field and laboratory experiments (Pederson, 1982; Hamilton, 1987), and the progression of biofouling can be divided into three phases (Characklis, 1981; Egan, 1987): a first phase of transport and adsorption of a wide range of periphytic bacteria; a second phase of growth and accumulation of adsorbed bacteria; and a third stationary phase. Recently, bacterial colonization of three substrata (stainless steel, aluminium, and polycarbon filters) was observed in an experimental continuous-flow seawater system at different periods of the year (Fera et al., 1989). The progression of areal bacterial density was analyzed. It was shown that early events in bacterial colonization of these materials may be affected by both water quality and type of substratum. These two parameters affect the cell adsorption phase (0 to 1 h), whereas seasonal variations in water quality were more important during the cell accumulation phase (1 h to 6 days) because of its effect on bacterial growth. Cell volume for the rod-shaped bacteria varied during the immersion time and seemed to be affected by seasons (Fig. 10).

COMPLEX MICROCOSMS

From Gnotobiotic Systems to Complex Microcosms

For studies of ecosystems and community levels of organization, the need to preserve at least part of the complex network of interrelated chemical, physical, and biological processes occurring within the ecosystem often leads to difficulties both in design of experiments and interpretation of results.

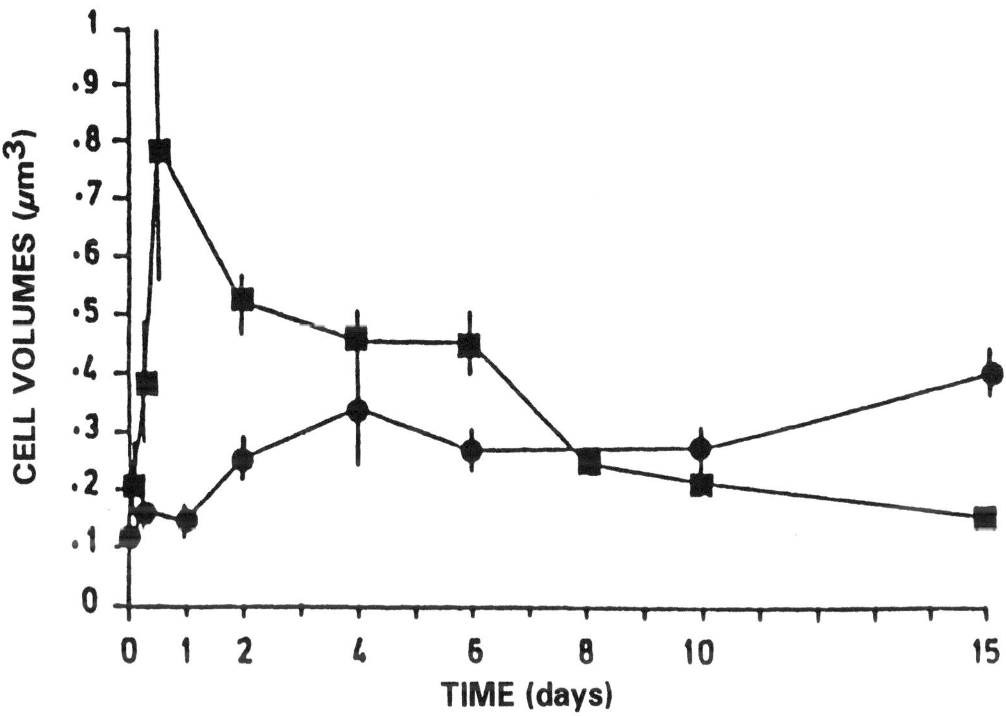

Figure 10. Development of bacterial films in an experimental flow system. Fixed rod-shaped bacterial volumes in summer (squares) and in winter (circles). From Fera et al. (1989).

Whittaker (1961) listed the methodological difficulties in studying his aquarium microcosms: "the inescapable complexity of aquarium processes, the changing character of these as the aquarium ages, and the interrelatedness of almost all that happens in the aquarium." Complex microcosms are, like mesocosms, portions of ecosystems or living models of ecosystems enclosed and maintained in controlled conditions. They include gnotobiotic microcosms, which are completely defined systems that include producers, consumers, and microorganisms so that their interactions can be observed. Examples include the gnotobiotic microcosms described by Taub (1969) and Taub and Crow (1980) for a freshwater community, and the synthetic microcosm described by Nixon (1969) which was a gnotobiotic microcosm of the brine shrimp *Artemia salina*. More complex microcosms have been based on transplanted portions of natural systems, with most or all of the species included. Pilson and Nixon (1980) have listed representative marine "microcosms" (*lato sensu*: including "mesocosms"). Their list includes a spectrum of sizes ranging from

19-l glass carboys to the 1,300-m^3 bags of CEPEX and the 1,000-m^3 containers of MERL. They quote that the common feeling in setting out to build enclosures, at least those thought to enclose pieces of real nature, seems intuitively to be that "bigger is better". They recognize, however, that "an experiment with systems containing four trophic levels might be accommodated in 1 or 2 cubic centimeters of sediment, with microbes through meiofauna, or in several hundred cubic kilometers of oceanic water, with diatoms through anchovies to squid and pilot whales." It is noteworthy that for studies of the theory of interactions in multispecies communities, the small microcosms would seem to be more appropriate. A big enclosure is not a panacea; a proper size should be clearly established for the purpose of the study (see Parsons, 1982).

Upwelling Event Simulation on Phytoplankton and Bacterial Communities

A 60-l pyramid-shaped microcosm was filled with recently upwelled surface sea water from the Southern Benguela upwelling system to follow the complex successions and adaptive responses of the bacterial community associated with phytoplankton growth and decay (Painting et al., 1989). The water was filtered (60 μm) to remove mesozooplankton and incubated under a diurnal light regime at *in situ* temperature (12°C) for 43 days. Particulate material in the microcosm was prevented from sinking by agitation with a stirring paddle. Growth or settlement of particles or microorganisms on the sides of the microcosm was prevented by a fine "bubble-curtain" of sterile air. Bacterial uptake of 14C-labelled substrate, incorporation of (methyl-3H) thymidine, and bacterial growth were followed in conjunction with changes in substrate sources during phytoplankton growth and decay.

The phytoplankton production and biomass, particulate carbon, and the development time of the bloom in the microcosm were within the same range as *in situ* values. Bacterial production rates associated with phytoplankton growth were similar to those recorded for the Southern Benguela upwelling region. These microcosm data seemed to be a realistic simulation of natural conditions; they suggested that different bacteria have specific substrate preferences, which probably confer competitive advantages under differing environmental conditions. The data indicated that bacterial substrate specificities, possibly in conjunction with flagellate predation, may be a significant mechanism in the control of microbial successions.

Effects of Turbulence on the Composition of Phytoplankton

Margalef (1963, 1967, 1978) was one of the first to stress the importance of turbulence as one of the most important inputs of external energy that probably controls phytoplankton assemblages. An experimental approach was proposed by Margalef (1963) to study the effect of different conditions of water mixing on phytoplankton composition; he used plexiglass, cylindrical, culture vessels, 2 m high and 15 cm in diameter, which were stirred and subjected to vertical gradients of temperature and light. Margalef's statement was that "no very important experimental results have been obtained, but the approach seems promising."

In the large 13-m^3 mesocosms at MERL, University of Rhode Island, a mechanical mixing device was included to experimentally analyze the influence of turbulent mixing on the ecology of a pelagic system. Nixon et al. (1980) demonstrated that the ecology of the system was significantly influenced by the turbulent energy dissipation rate. Their results suggested that, in both mesocosms and the natural system (Narragansett Bay), the average rate of biological energy dissipation seemed to be greater than that dissipated by the physical system in turbulent water mixing.

It has been interesting to examine the effects of different levels of small-scale turbulent mixing on marine phytoplankton populations in microcosms. An advanced modification of the Margalef tubes was developed by Estrada et al. (1987). The 30-dm^3 cylindrical microcosms (Fig. 11) were maintained at constant temperature and under a 12 h light cycle. A complex set of combinations using different oscillating grids and different oscillation frequencies from 20 to 70 min^{-1} was applied to each pair of experimental tubes. Although stirring and turbulence refer to different concepts, it was assumed that increased stirring of the medium produced increasing levels of turbulent mixing. Experiments were started at different times of the year (June, October) by filling vessels with sea water from Masuon Harbor, north of Barcelona. Filtration through a 150-μm mesh excluded most adult zooplankters, whereas phytoplankton, larval zooplankton, and ciliates passed through.

Statistical analyses of the phytoplankton inventories were based on the different experimental treatments: (1) effects of stirring on chlorophyll concentration and cell numbers; (2) effects of stirring on phytoplankton composition; and (3) influence of the initial population. The confinement of phytoplankton populations in the experimental tubes triggered a sequence of events depending on the initial conditions and modulated in a reproducible way by the treatment applied to the microcosms. As reported in other types of enclosures (Parsons et al., 1978), an initial bloom of fast-growing centric diatoms was observed in combination with a pronounced increase in chlorophyll in both unstirred and stirred tubes. These initial populations consumed nutrients, sank to the bottom, and were replaced by coccolithophorids, dinoflagellates, and flagellates. The stirring treatments modified the timing and intensity of the initial population peak (which occurred independently of the stirring and under a variety of light intensities or predator concentrations) and, within each experiment, induced reproducible changes in the composition of the phytoplankton assemblages. This design allowed experimentation with small-scale turbulent motions which may have important implications in relation to successions of phytoplankton populations in sequences ranging from diatoms dominating in turbulent and fertile water, to populations found in exhausted and stratified environments (typically dinoflagellates; see Margalef, 1978).

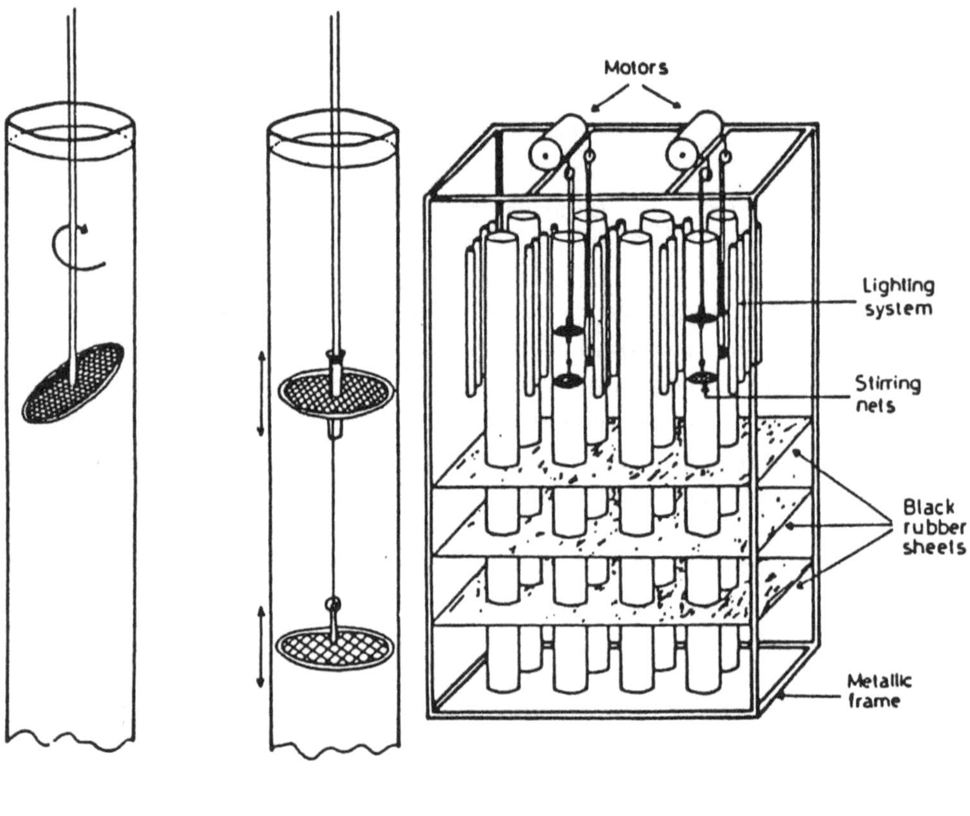

Figure 11. Cylindrical microcosms to study the effects of turbulence on the composition of phytoplankton assemblages. Right, a set of eight microcosms in a metallic frame; center, oscillating grids; left, rotating grid. From Estrada et al. (1987).

Marine Sediment Microcosms

Detritus Decomposition

Various experimental approaches have been developed in the last decade to better understand the role played by detritus in the rate of changes in total mass and various biochemical pools, including nitrogen. In a recent and well-documented review, Tenore (1987) showed that a point of caution is needed in interpreting data in the literature on changes in nitrogen content of aging detritus. Experiments in stagnant flasks can produce several artifacts, such as decomposition being controlled by the culture-medium nutrient concentration and/or by metabolic products that inhibit microbial activity. Moreover, flow-through microcosms and *in situ* bags can be affected by the intensity of mixing. Tenore (1987) gives three examples of microcosm studies: (1) where mixing was minimized (Tenore et al., 1984) so that rates of decomposition were kept at a minimum (e.g. for *Spartina*, a total mass loss of only about 20% in 280 days); (2) the vigorously mixed carboys used by Rice and Hanson (1984) in which total mass loss of *Spartina* detritus was about 14% for 26 days; and (3) the *in situ* bags studied by Valiella et al. (1984) in which the total weight loss was about 80% after 300 days. Hanson and Tenore (1981) found that aged detritus kept anaerobically displayed a positive correlation between increasing rate of loss of detrital mass and the intensity of mixing.

Benthic Mineralization and Metabolism

The decay rates of particulate organic carbon (POC) and nitrogen (PON) were followed during 94 days in microcosms consisting of sieved surface sediment in 4.5-cm diameter plexiglass cores (Kristensen and Blackburn, 1987). These microcosms were supplied with overlying sea water and incubated at 22°C to measure solute exchange between the sediment and the overlying water. Three treatments were applied (Fig. 12): (1) aerobic conditions with a natural density of *Nereis* polychaetes; (2) aerobic conditions without *Nereis*; and (3) anaerobic without polychaetes. The presence of *Nereis* increased the net decomposition of organic-N and organic-C by 1.6 and 2.6 times, respectively, relative to the aerobic microcosm without polychaetes (losses expressed as percentage of the starting material).

Mineralization was maximal in microcosms containing *Nereis*. Surprisingly, the next greatest losses were measured in anaerobic microcosms. The C:N ratio of the mineralized material was always less than the original detritus (Fig. 13). At the end of the 94-day incubation, the material undergoing mineralization in the *Nereis* aerobic microcosms had a C:N ratio of 16, versus 4 in both the aerobic and anaerobic microcosms without animals. This result implies that the rate of mineralization can be significantly increased by macrofaunal digestion of the input material.

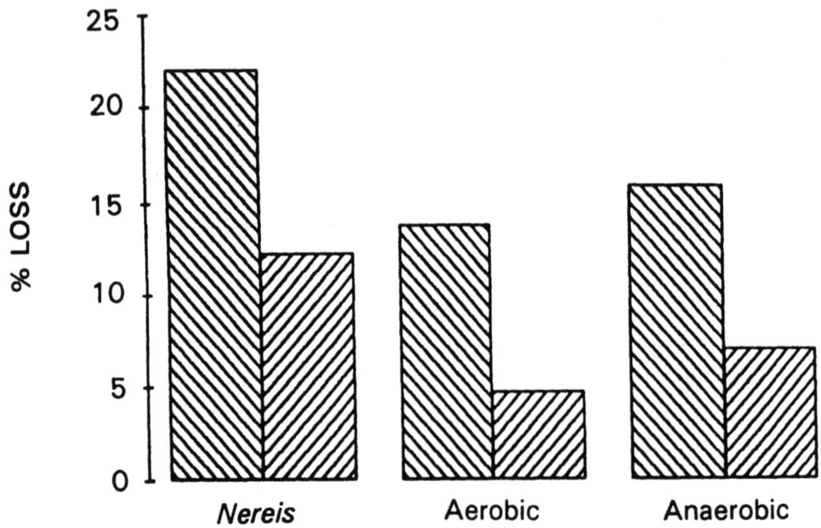

Figure 12. Mineralization of detritus in microcosms, expressed as percentage of organic matter (first column) and organic carbon (second column) after 94 days of experiment: three microcosms with *Nereis virens*; three used as aerobic controls without *Nereis*; and three supplied with an anoxic (N_2-purged) water phase. From Kristensen and Blackburn, 1987, modified in Blackburn (1987).

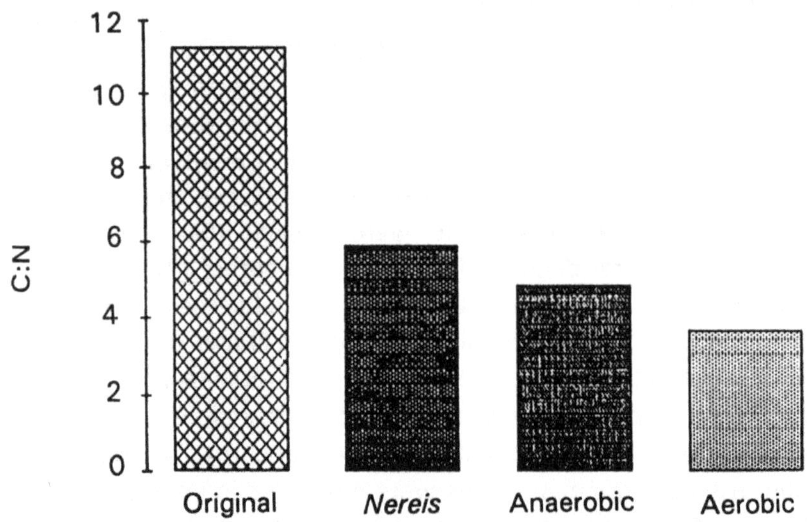

Figure 13. C:N ratios of the organic matter that had been mineralized in *Nereis*, anaerobic, and aerobic microcosms during the 94-day experimental period. The first column represents the original sediment. From Kristensen and Blackburn, 1987, modified in Blackburn (1987).

The ventilatory pumping of *Nereis* increased the oxygen flux, but most of the sediment in microcosms with worms appeared to be anoxic at the end of the experiment (96 days). The presence of macrofauna increased the rates of degradation by some mechanism other than by the direct effect of oxygen. Blackburn (1987) concluded that "the rate of mineralization is due to macrofaunal digestion of the input material, while in others it may be due to the mechanical exposure of fresh surfaces to bacterial colonization." Both phenomena are probably combined in nature. The results obtained from these microcosm experiments cannot be extrapolated directly to *in situ* conditions, because natural heterogeneity has been suppressed by the homogenization of sediment samples. Nevertheless, the measurements can be used for comparative investigations and "should merely be regarded as idealized sediments without natural heterogeneity" (Kristensen and Blackburn, 1987).

In a recent microcosm study, Andersen and Kristensen (1988) showed that benthic macrofauna have an important impact on estuarine benthic community metabolism. This study was developed in experimental cores kept in either a light/dark cycle (autotrophic condition) or in continuous darkness (heterotrophic condition) over 65 days. The purpose was to evaluate the effects of benthic fauna and light on fluxes of O_2, CO_2 and DIN (NH_4^+ + NO_2^+ + NO_3^+) by following the interactions between three macrofaunal species (*Corophium volutator, Hydrobia* sp., and *Nereis virens*) and benthic microalgae. Ten microcosms (two of each type including two controls with no animals) were exposed to a 12-h light and 12-h dark cycle (L/D-microcosms) and ten were kept in continuous darkness (D-microcosms). Sea water overlying the sediment cores was continuously renewed at a rate of approximately 50 ml/minute. After 10 days, the benthic primary production in L/D-microcosms stabilized O_2 and CO_2 flux rates; on the contrary, D-microcosms displayed decreasing O_2 and CO_2 flux. Community respiration was stimulated in darkness for O_2 and CO_2. Highest rates of DIN were measured in microcosms containing animals, mainly because of an increased NH_4^+ flux. One interesting implication of these experiments is that dark O_2 uptake, CO_2 and DIN fluxes were in the range reported in the literature for coastal sediments (Hargrave and Phillips, 1981; Andersen and Helder, 1987).

Differential effects of *Corophium, Hydrobia,* and *Nereis* were found. The deep-burrowing *Nereis virens* increased the dark fluxes; however, this polychaete contributed only 10% of the increase in oxygen uptake, indicating a strong stimulation of microbial activity in the sediment. The effects of *Corophium* and *Hydrobia* seemed to be less than that of *Nereis*. A "filter" effect of benthic microalgae, i.e. the decrease of NH_4^+ from sediments in light as described by Henriksen et al. (1980), was also detected. This phenomenon, providing a short-cut in the nitrogen cycle, should be considered for measurements of DIN fluxes. A dark "filter" effect was found in light-day microcosms after 8 h of darkness; this phenomenon counteracts the nitrogen export through ammonia production of the fauna. Furthermore, these experiments showed that benthic macrofauna may be responsible for an active removal of nitrogen form the ecosystem by stimulating the denitrification process.

Effects of Macrofaunal Excretion and Turbulence on Nutrient Dynamics

As in planktonic studies, the utilization of intermediate-sized microcosms (between culture vessels of a few milliliters and mesocosms) allows measurements on the important question of the comparative effects of macrofaunal excretion and physical turbulence on nutrient dynamics at the seawater-sediment interface.

The respective effects of macrofauna excretion and turbulence on the NH_4^+ and NO_3^- fluxes occurring at the seawater-sediment interface were studied using a batch of eight closed-circuit microcosms, with sea water percolating through sediment from an oyster bed located in the Bay of Morlaix (Fig. 14). Each microcosm consisted of a bicompartmentalized PVC box: in compartment A, a plastic grid covered by a 100 μm nitex screen was fixed at 6 cm from the bottom; compartment B had a regular bottom. A communication between the compartments was maintained over compartment A giving a total sediment layer of 400 cm^2. The microcosms were kept under a natural photoperiod, and short-term experiments (7 h) were performed in winter and in summer. Microcosms were set up 2 weeks prior to the experiments as this delay was estimated to be appropriate for the recovery of the sediment layer and the seawater-sediment interface.

The two epibenthic macrofaunal species studied were characteristic of oyster beds: the oyster *Crassostrea gigas* (a filter-feeder) and the crab *Carcinus maenas* (a scavenger). In the presence of macrofauna, the NH_4^+ content (measured in the water column and the pore-water) was markedly increased, with NH_4^+ fluxes estimated at the water-sediment interface of 1 to 2 orders of magnitude greater than in the absence of macrofauna during both seasons (winter and summer).

Turbulence simulation in the water column led to a simultaneous increase of NH_4^+ in the water column and a decrease of NH_4^+ in the pore-water (Regnault et al., 1988). These nutrient exchanges likely resulted from a physical rather than a biological process. Conversely, NO_3^- content was not significantly modified by either turbulence or macrofauna. Ammonia excretion rates of crabs and oysters were influenced by turbulence with differences according to the season and the considered species. In the control (no animals), net fluxes of NH_4^+ at the seawater-sediment interface were similar in winter and in summer (with values of 79.9 and 79.2 μM m^{-2} h^{-1}, respectively). In the presence of crabs and oysters, the net flux of NH_4^+ was twice as high in winter, whereas the ammonia flux was not significantly increased in summer. In winter, turbulence markedly stimulated the net flux of ammonia in the presence of macrofauna, whereas this excretion was reduced to a very low or negligible value in summer. Concentrations of NH_4^+ and NO_3^- were of the same order of magnitude as in nature (Regnault et al., 1988).

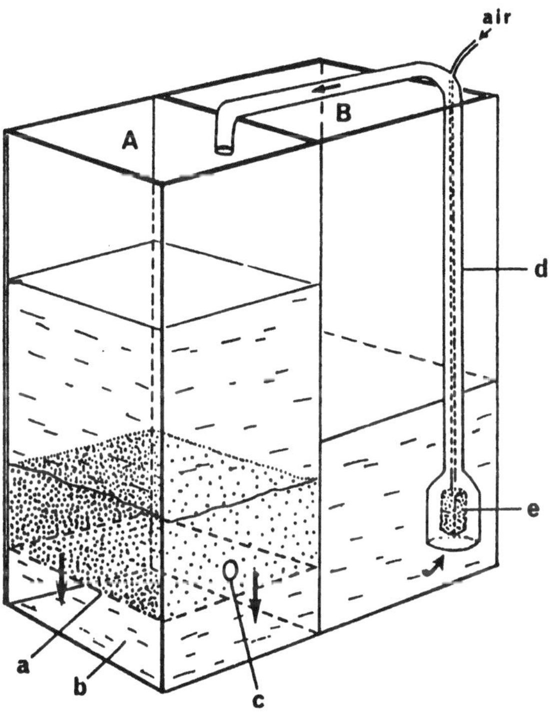

Figure 14. A closed-circuit microcosm, with sea water percolating through sediment to study the effects of macrofauna excretion and turbulence at the seawater-sediment interface. Sediment is in compartment A. a, plastic grid with nitex screen; b, sea water percolating from the sediment; c, hole for communication between A and B; d, air-lift system; e, air stone; arrows for water circulation. From Regnault (1986).

Microcalorimetric Investigations on Microbial Ecology

The application of microcalorimetry in the field of microbial ecology is recent. New microcalorimeters are highly sensitive and reliable. They have been used recently for qualitative and quantitative characterization of energetic changes occurring in terrestrial soils and in freshwater and marine sediments (Ljungholm et al., 1979; Lasserre, 1980, 1984; Pamatmat et al., 1981; Pamatmat, 1982, 1984; Gustafsson, 1987).

A great advantage of microcalorimetry, compared with other techniques used for measuring metabolism, is its nonspecificity. Because any chemical or physical process will be accompanied by an enthalpy change, the method is completely general. It might seem that there would be considerable difficulty in interpreting results, because resolution of the processes taking place would become very complex in living organisms. However, the practical situation is not usually so complex as it seems at first sight.

Pamatmat (1982) presented convincing results for the heat production rate of sediment samples collected from a San Francisco Bay mud flat. Measurements were performed during aerobic and anaerobic exposures. The decreasing specific rate of metabolic heat production of sediments, owing to oxygen deprivation in the calorimeter vessel, was almost overcome by using a perfusion vessel instead of closed ampoules (Gustafsson and Gustafsson, 1985).

Lasserre (1980, 1984) and Lasserre and Tournié (1984) developed a flow microcalorimetric system to characterize the metabolic heat flux occurring at the seawater-sediment interface of microcosms representing different estuarine and littoral ecosystems. The 35-ml microcosms consisted of glass chambers loaded with 10 ml of sieved sediment and 25 ml of sea water. These microcosms were maintained at 19°C with a daylight cycle of 6 hours. Sea water 5 mm above the sediment surface was pumped through a filter at a constant flow rate of 23 ml hour^{-1}. This 'circulating interface' was taken through a flow microcalorimeter and an oxygen electrode before returning to the microcosm (Fig. 15). During experiments, microcosms were maintained in darkness to eliminate photosynthesis. The technique provided a reliable method by which changes in the metabolic activity at the seawater-sediment interface could be detected and quantified. The underlying metabolic processes have been identified by adding ^{14}C-glucose to the circulating interface and measuring the ATP pool level. Moreover, the microcosms were submitted to experimental disturbances by introducing known quantities of organic matter or by bioperturbation at the sediment surface through the actions of small deposit-feeders (*Hydrobia*) and meiofauna (copepods).

The power-time curves (PTCs) obtained with experimentally enriched microcosms are highly reproducible. The microcalorimeter thus acts as a highly sensitive microwatt-meter having a limit of detectability of 0.5 microwatt. With the circulation microcalorimetric system, a density of 10^5 bacteria gave a routine lower metabolic heat production of 1 microwatt. After nitrogen enrichment, the exponential transitory phase of the PTC was clearly correlated with aerobic glucose metabolism, ATP pool level, and oxygen tension. The microevents noted on the PTC during the exponential phase were also observed on the oxygen-time curve, and a linear correlation was observed between the heat production and oxygen utilization. Calorimetric

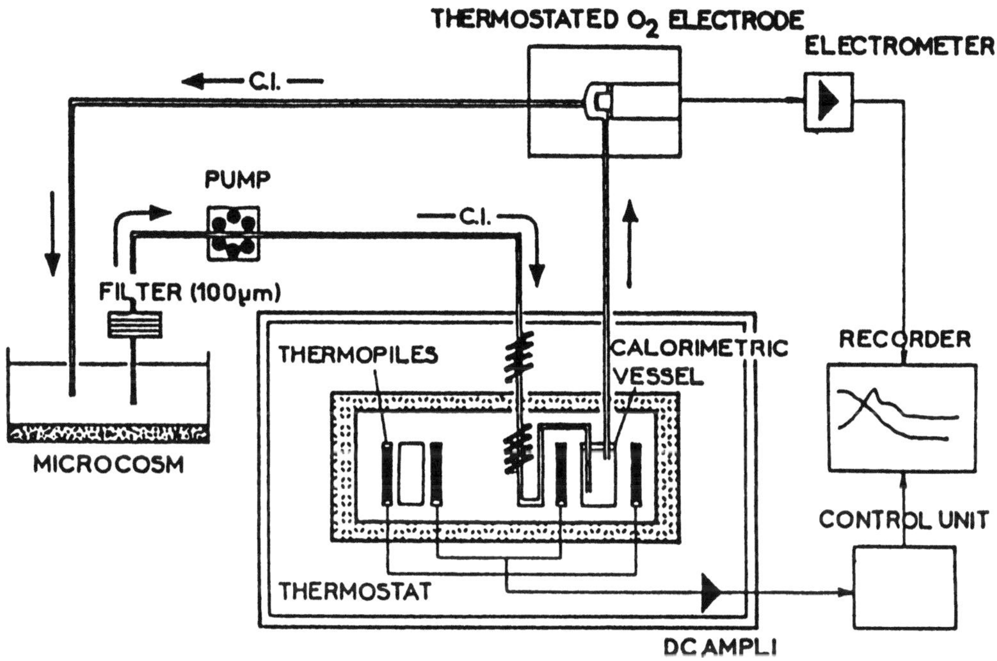

Figure 15. A flow microcalorimetric and oximetric system to characterize the metabolic heat flux and oxygen tension occurring at the water-sediment interface. CI, circulating interface. From Lasserre, 1980, modified in Lasserre and Tournié (1984).

and oximetric responses showed alternate periods of homogeneity (summer and winter months) and of heterogeneity (spring and autumn months): periods of homogeneity were characterized by very reproducible PTCs and OTCs; periods of heterogeneity displayed transition stages between summer and winter types (Tournié and Lasserre, 1984). The spring heterogeneous PTCs obtained from March to June can be interpreted as a gradual evolution from sigmoid (winter type) to unimodal (summer type) PTCs (Fig. 16).

Although many problems remain to be solved, the results indicated that direct microcalorimetry may provide new information on the overall temporal behavior of complex mixtures of microorganisms colonizing the seawater-sediment interface of coastal environments. Furthermore, the same microcalorimetric approach proved well suited for the study of the effects of bioturbation of meiofauna and small deposit-feeders.

Figure 17 shows the heat dissipation produced at the water-sediment interface by the introduction of the meio-epibenthic copepod, *Eurytemora hirundoides*, at a density of 2 individuals per centimeter. These experiments showed that copepods induced a metabolic perturbation in the overlying circulating sea water; this perturbation lasted a few hours, and decreased until a new steady-state plateau was reached, i.e. the same level as before the copepod introduction. In the presence of the small prosobranch *Hydrobia ventrosa* (2 individuals cm^{-2}), heat dissipation at the water-sediment interface was dependent on the preincubation time of the microcosms (2, 7, 13, and 16 days, respectively; see Fig. 11B).

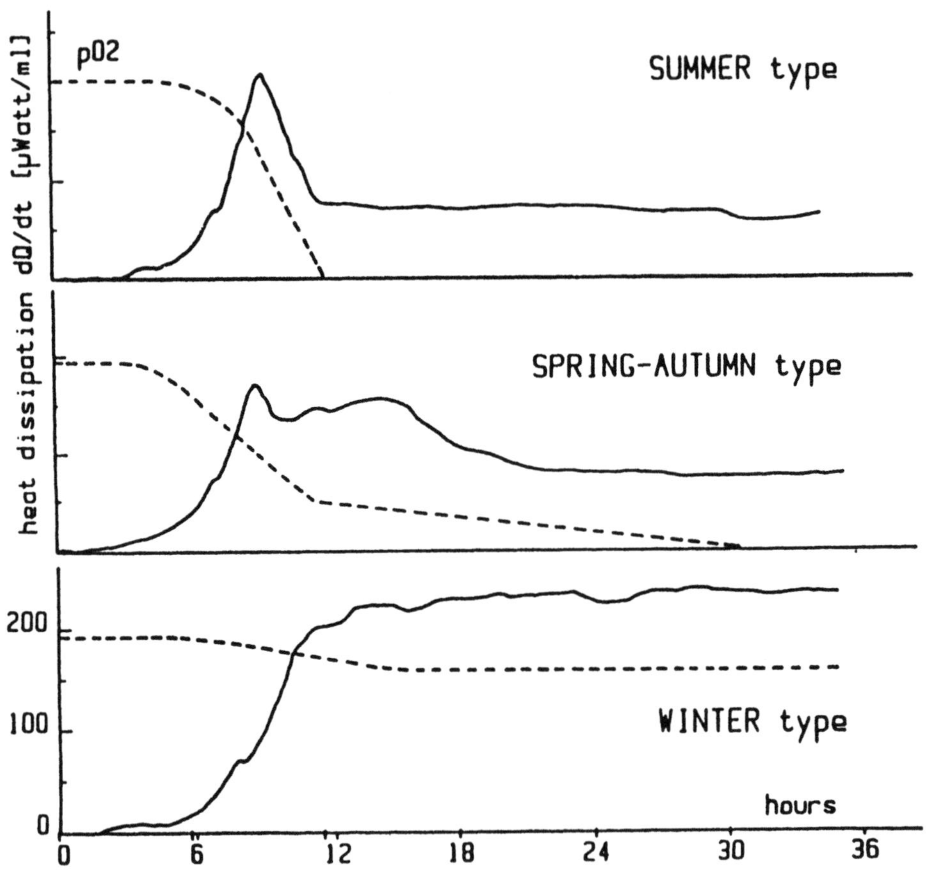

Figure 16. Typical seasonal microcalorimetric and oximetric patterns produced in microcosms at the seawater-sediment interface after an experimental eutrophication. A, unimodal summer PTC; B, bimodal spring-autumn PTC; C, sigmoid winter PTC, with their respective oxygen-time curves (pO_2). From Lasserre et al. (1986).

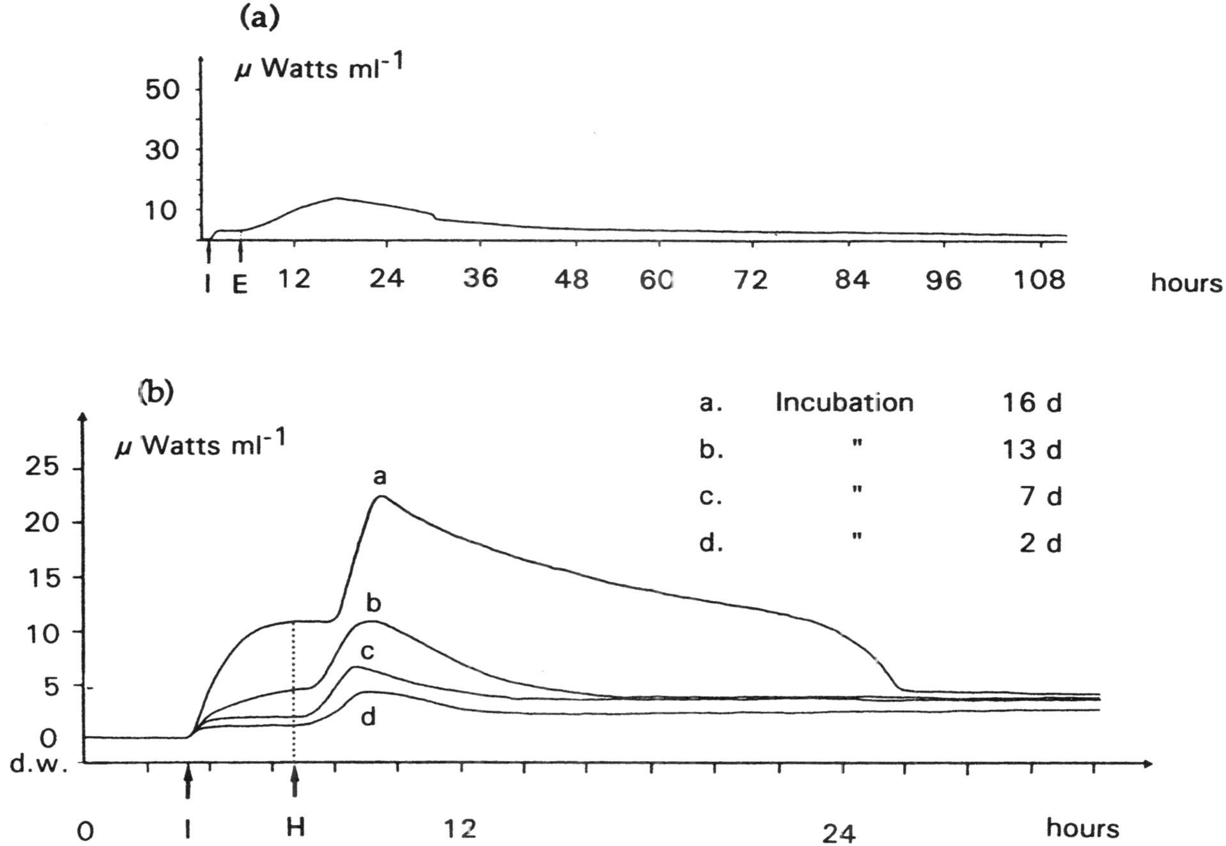

Figure 17. Microbial heat dissipation at the seawater-sediment interface generated by the introduction of estuarine epibenthic feeders: the copepod ***Eurytemora hirundoides*** (a) and the gastropod *Hydrobia ventrosa* (b) in different microcosms. In (b), the microcosms have been "aged" 2, 7, 13, and 16 days, respectively. H: *Hydrobia*. Modified from Lasserre (1980).

The mechanisms underlying PTC evolution were due to the rapid (hours) succession of bacterial populations with different dynamics and physiological capabilities (Lasserre et al., 1986). "Summer" type microcalorimetric curves were associated with bacterial populations responding to eutrophication by rapid and well marked successions, ending in the selection of a community dominated by anaerobes. These anaerobes were characterized by very high catabolic potentialities. During the first part of the microcalorimetric curve, from time zero (PO) to the PTC peak (P3), aerobes gradually increased in specialization (Fig. 18). When the "winter" type calorimetric curve was observed, the bacterial successions were less marked. From the beginning of the experiment to the secondary steady state, the populations remained very similar and their catabolic potentialities were identical. As a consequence, the specific rate of heat was significantly higher than in summer.

More generally, these studies offer an example of the value of direct microcalorimetry in studying adaptive properties of ecosystems. Here, heat dissipation with respect to seasons is an adequate parameter of the ability of the system to return, after transitory oscillations, to a new steady state (a parameter of ecological "resilience"). Mixed bacterial populations displayed different energetic regulations of adaptive nature. A characteristic feature was the increase in the specific heat production rate shortly after the perturbation (nutrient enrichment), followed by a decrease that probably indicates subsequent functional adaptational changes in bacterial populations, and finally a stationary state. The stationary phase indicated a maximum efficiency with probable adaptation of enzymatic systems to new environmental conditions.

An intriguing problem is to understand how these successional events relate to the minimum entropy production based on linear irreversible thermodynamics (Prigogine-Wiame, 1946; Schneider, 1988). Recent experimental developments have shown that the thermodynamics of nonequilibrium processes can be used to deduce quantitative relationships and equations of importance to developmental biology (Lamprecht and Zotin, 1978). For Zotin (1985), "the possibility is not to be excluded that thermodynamics might be used to obtain phenomenological equations appropriate to the theory of ecosystems."

Trophodynamic Studies

In studies of food webs, a variety of measurements are made of population dynamics and of physiology. For population dynamics, rates of reproduction, growth, age, and size of populations can be obtained *in situ* by exclusion experiments with cages (e.g. Stephenson and Searles, 1960) and by introducing experimental enclosures such as bell jars or flow-through systems

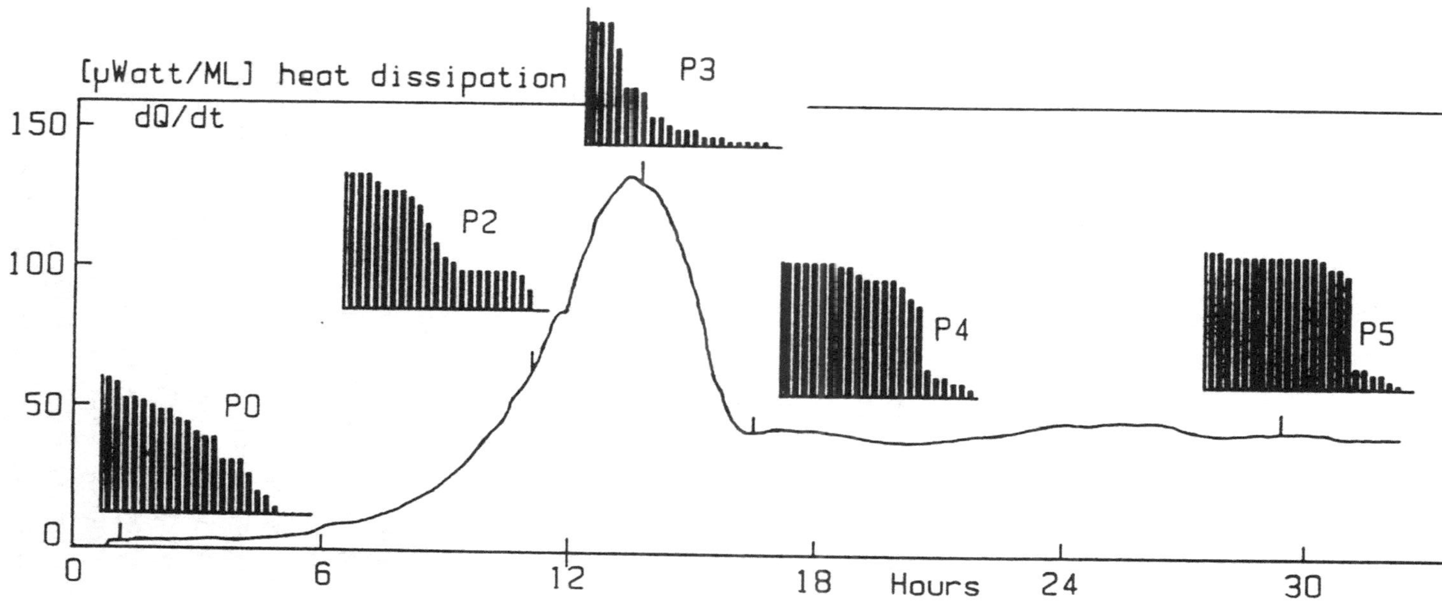

Figure 18. Evolution of microbial catabolic potentialities during a "summer" type microcalorimetric response to eutrophication. The five bar charts (P0 to P5) characterize the specialization of the bacterial population, expressed as the ability to metabolize a variety of organic substrates. From Lasserre et al. (1986).

for field experiments; these produce meaningful results, but the techniques are laborious and time-consuming.

Laboratory experiments, generally short-term, are conducted to measure such physiological functions as feeding, respiration, excretion, and other metabolic processes. Such short-term experiments give instantaneous budgets of metabolic processes of individual animals or plants which can furnish quantitative estimates required by ecosystem models. Extrapolation to populations over their entire life cycle is not so obvious.

Laboratory Feeding Experiments

Interesting variations of chemostats for continuous culture of planktonic and bentho-pelagic organisms have been developed recently. These microcosms employ either naturally occurring foods or food mixtures of known composition.

Nutrition and growth of planktonic herbivores and ciliate protozoa have been studied in chemostat steady-state conditions (Droop, 1976; Droop and Scott, 1978; Fenchel, 1977; Scott, 1980, 1985). Experiments developed in microcosms with homogeneously, ^{14}C-labelled, hay particles inoculated with bacteria and various protozoan grazers of bacteria (Fenchel, 1977) have demonstrated that protozoan grazing of bacteria has a significant effect, not only in controlling bacterial population size, but also in controlling metabolic activity as measured by the breakdown of structural carbohydrates of plant tissue, and in the cycling of mineral nutrients.

A marine ciliate, *Strombidium* sp., was grown bacteria-free in a chemostat continuous-culture system (Fig. 19) and fed on the alga *Pavlova lutheri*. The advantage of the chemostat was to maintain a homogeneous population, in which rate variables were really obtained from state variables (Scott, 1985). The growth rate of the ciliate was maintained constant whilst the growth rate of the alga was varied, the latter producing little change in its energy content. The specific filtration and ingestion rates of the ciliate were measured: the efficiency of ingestion remained constant throughout the experiment with a mean of 91%; the overall growth efficiency also remained constant, but with a mean of only 6.5%. Compared with the results obtained by Fenchel (1980), the maximum filtration rate (in body volumes per hour) and the specific rates (based on the mean ciliate body volume) were not as great. Considering the low growth efficiency of 6.5%, the ciliate must be continuously dissipating 93.5% of its potential energy (!) (Scott, 1985).

Droop (1976) described continuous-culture systems suitable for mariculture: a cheap 4 x 15-l system for studying algal nutrition and growth kinetics of algae (e.g. of the green alga *Branchiomonas submarina*) and an apparatus for culturing aquatic invertebrates (e.g. the marine rotifer

Brachionus) under bacteria-free conditions. Trials with the chemostat showed that the *Brachionus* - *Branchiomonas* system would prove a good model of an algal-herbivore link.

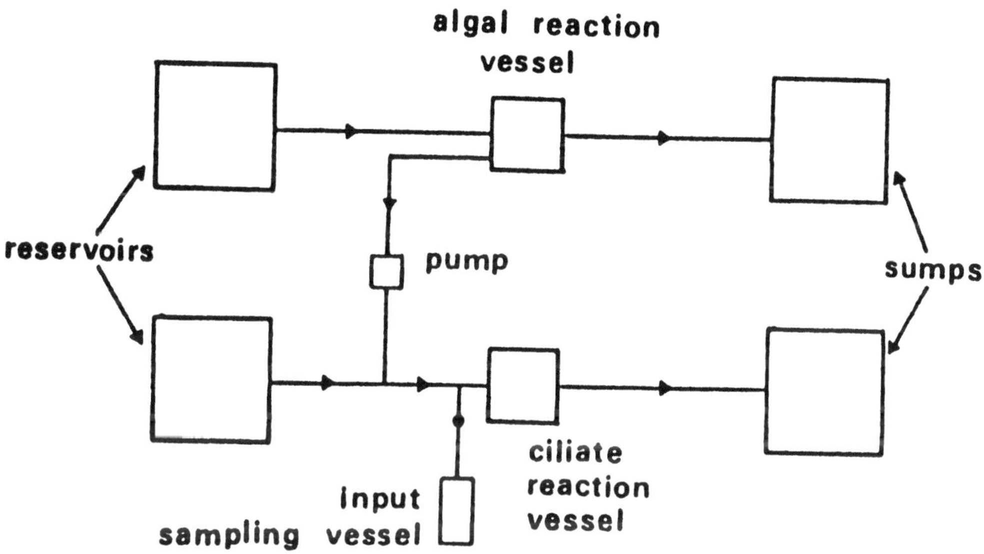

Figure 19. Diagrammatic view of the predator-prey chemostat system used for the marine ciliate *Strombidium* fed on the alga *Pavlova*. From Scott (1985).

Apparatus for the continuous culture of pelagic crustaceans have been described (e.g. Paffenhöfer, 1976). The apparatus described by Ringelberg (1977) for the continuous culture of *Daphnia magna* involved an algal culture which sustained a population of 400 to 700 *Daphnia*. The microcosm showed a strong tendency for self-maintenance despite fluctuations in the algal and zooplankton populations.

Parsons and Bawden (1979) described a similar, two-step, food chain for the continuous culture of a bentho-pelagic amphipod, *Anisogammarus pugettensis* , that was maintained for over 6 months. They observed, during this period, three generations of amphipods when food was supplied in excess. The purpose of the experiment was to follow growth parameters under conditions of limited food supply. The apparatus (Fig. 20) consisted of a growth chamber for the diatom *Amphiprora paludosa* which was easily consumed by the amphipods. The amphipod cages retained the animals and algae, but allowed free passage of sea water. Food was supplied at rates controlled by the pumping velocity. Fifty newly-hatched amphipods were

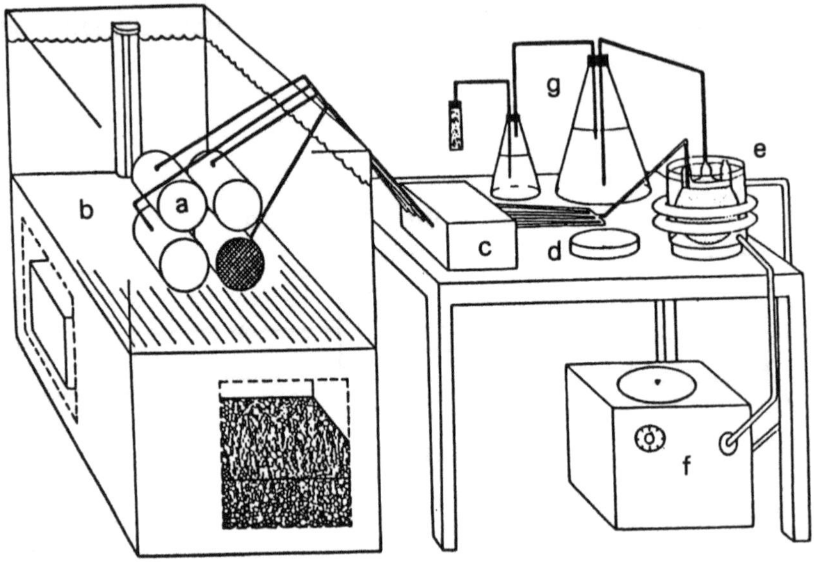

Figure 20. Microcosm for the continuous culture of a bentho-pelagic amphipod. a, Amphipod cages; b, aquarium; c, proportionating pump; d, magnetic stirrer; e, phytoplankton culture; f, temperature-controlled bath; g, nutrient medium for phytoplankton. From Parsons and Bawden (1979).

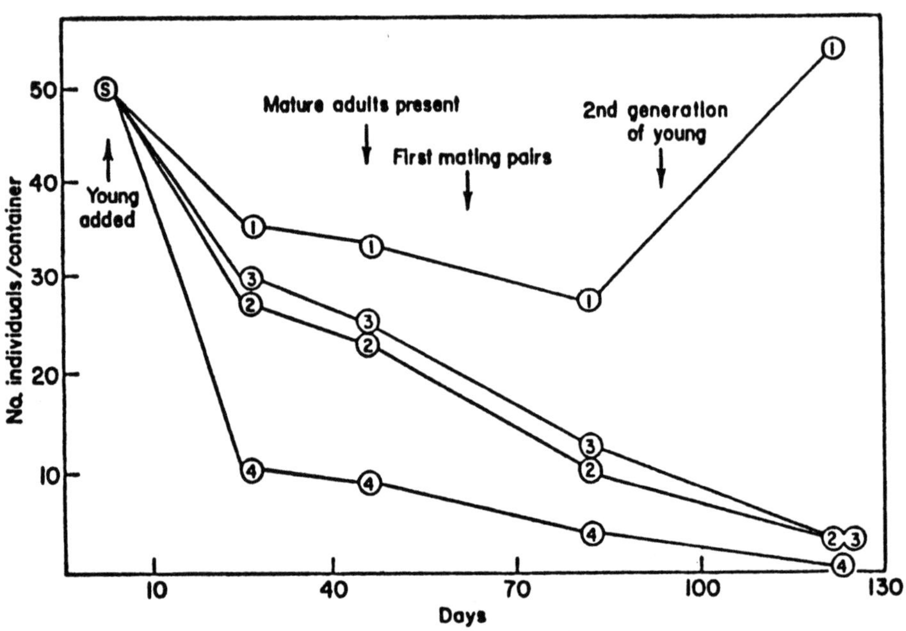

Figure 21. Numerical changes of amphipods in cages 1 to 4. From Parsons and Bawden (1979).

placed in each cylinder, and the changes in amphipod numbers over the course of the experiment are shown in Fig. 21. In container 1, a cyclical fluctuation seemed to have commenced. In containers 2, 3, and 4, however, amphipods were not able to maintain population densities. Although growth of individual amphipods was observed in containers 2, 3, and 4, the food supply was insufficient to support the populations in terms of the animal's life cycle. Food maintenance requirements for amphipod populations were found to be higher than those calculated from respiration measurements.

Meiofauna-Microorganisms Interaction

The partitioning of food resources between microorganisms and benthic meio- and macrofauna has been studied extensively in experimental marine sediment microcosms. In columns of sand simulating beach conditions (McIntyre et al., 1970), it seemed that meiofauna fed mainly on bacteria growing on soluble organic matter and that, within the sand, there were meiobenthic predators such as turbellarians feeding on harpacticoid copepods, so that all the energy entering the sandy interstitial ecosystem was dissipated within the system. A marine sand ecosystem of 360 liters was maintained for 28 months (Boucher and Chamroux, 1976; Chamroux et al., 1977) with sea water percolating through the sediment. It received regular additions of soluble amino acids. The aim was to control bacterial development in order to obtain a population that might serve as food for meiofauna without causing imbalance to the medium in this complex environment. Bacteria and nematode populations maintained densities similar to *in situ* levels over several months. Nevertheless, copepods fluctuated dramatically. As in the study by McIntyre et al. (1970), it was difficult to maintain the original natural heterogeneity over several months or a year. Mesocosms are probably better for maintaining a more natural spatial-temporal heterogeneity over long periods.

Epidemiology and Ecology of *Vibrio cholerae*

Laboratory microcosms were recently employed to evaluate the influence of selected environmental parameters, such as organic nutrient concentration, pH, and salinity, on the growth and survival of a toxigenic strain of *Vibrio cholerae* that is associated with the copepods *Acartia tonsa* and *Eurytemora* sp. (Huq et al., 1984). Approximately 500 copepods were placed in 500 ml of filter-sterilized water in 2-l flasks. Cells of *Vibrio* were added to each flask so that the final cell concentration in each microcosm was about 10^4 CFU per milliliter. By increasing temperature up to 30°C, a significant effect on the multiplication of *V. cholerae* was demonstrated, as was attachment of the cells to living copepods. Moreover, maximum growth of *V. cholerae* and attachment to copepods occurred at 15°/oo and at pH 8.5. An acidic pH (6.5) produced a rapid decline in plate counts of *V. cholerae*, probably because of the early death of copepods; cell counts decreased in all experimental flasks when copepods in those microcosms began to die.

The conditions affecting attachment of *Vibrio cholerae* to laboratory-cultured copepods are important in clarifying the natural ecological relationship between *Vibrio* and planktonic copepods, as this is important in the epidemiology of cholera, for which *V. cholerae* serovar 0.1 is the causative agent. In temperate regions, the inability of *V. cholerae* to multiply and survive at low winter temperatures would reduce the risk, even if large populations of copepods were present. In warmer regions, such as Bangladesh, the summer water temperature often exceeds 25°C for prolonged periods and zooplankton blooms occur after the monsoons. Based on data collected in microcosm experiments, a combination of high water temperature and an abundance of copepods would be optimal for the survival and multiplication of *V. cholerae*, despite the low salinities (1 to 5°/oo) found in much of the delta regions of Bangladesh. The ingestion of unpurified water containing copepods heavily contaminated with *V. cholerae* may offer an inoculum of sufficient size to cause the disease in humans. Therefore, observations made in experimental microcosms designed to replicate conditions in nature aid in understanding the natural ecology of *V. cholerae*.

CONCLUSION

The experiments conducted so far with microcosms are important tools at the lower level of ecological organization, where one is dealing with bacteria, ciliates, and planktonic, meiobenthic, and small macrobenthic components. Microcosms are necessary adjuncts of experiments conducted with mesocosms, for the experimenters can control many of the boundary conditions and manipulate individual portions of subsystems at different scales of space and time. The data-sets in microcosms are not merely static measures but represent dynamic ecosystems, with small-scale and short-term behavior patterns that are difficult to quantify by other means.

Microcosms represent a transitional stage between the precision of laboratory single-species cultures and mesocosms exceeding 1 m^3 in volume. The continuum from the reductionist test-tube assay to both quasi-natural and semi-controlled micro-mesocosm conditions provides a means of reconciling the dichotomy between experiments and field surveys of natural ecosystems.

For Nixon et al. (1980), "It is intuitively more satisfying to study the ecology of coastal marine waters with large tanks or bags of sea water and marine sediments than to extrapolate from carboys, experimental salt marsh plots, or even small flasks of sea water." This remark is probably intuitively appropriate; nevertheless, microcosm studies have contributed to the development of a number of general ecological concepts, and recent results presented in this chapter show that some recently designed microcosms, when they are appropriately scaled, dynamically mimic natural conditions on a small scale.

The validity of microcosms as a simplification for studying the fates or effects of environmental perturbation or trace contaminants needs to be assessed. We suggest this be done by extensive field-mesocosm and mesocosm-microcosm comparisons.

An interesting value of microcosms is that they provide increasingly important information for refining and developing theoretical ecosystem ecology that suggests fundamental ecological/thermodynamic processes (Schneider, 1988). Microcosms enable experimentally controlled studies of functioning ecosystems with cybernetic or negative feedback loops in place. This knowledge is not only valuable in its own right but may be more useful in an indirect, rather than a direct, way in understanding the robustness of these systems and so in evaluating some general effects of stress (Steele, 1979). In carefully controlled situations, a microcosm may respond to stress very much like the natural marine ecosystem.

LITERATURE CITED

Andersen, F. O. and W. Helder. 1987. Comparison of oxygen microgradients, oxygen flux rates and electron transport system activity in coastal marine sediments. *Mar. Ecol. Prog. Ser.* **37**: 259-264.

Andersen, J. O. and E. Kristensen. 1988. The influence of macrofauna on estuarine benthic community metabolism: a microcosm study. *Mar. Biol.* **99**: 591-603.

Beyers, R. J. 1963. The metabolism of twelve aquatic laboratory microecosystems. *Ecol. Monogr.* **33**: 281-306.

Blackburn, T. H. 1987. Microbial food webs in sediments. Pp. 39-58. **In**: M. A. Sleigh [ed.], *Microbes in the Sea*. Chichester: Ellis Horwood, John Wiley.

Blackburn, T. H. and J. Sorensen [eds.]. 1988. *Nitrogen Cycling in Coastal Marine Environments.* Chichester: John Wiley. 451 pp.

Boucher, G. and S. Chamroux. 1976. Bacteria and meiofauna in an experimental sand ecosystem. I. Material and preliminary results. *J. Exp. Mar. Biol. Ecol.* **24**: 237-249.

Burkill, P. H. 1987. Analytical flow cytometry and its application to marine microbial ecology. Pp. 139-166. **In**: M. A. Sleigh [ed.], *Microbes in the Sea*. Chichester: Ellis Horwood, John Wiley.

Chamroux, S., G. Boucher, and P. Bodin. 1977. Etude expérimentale d'un écosystème sableux. II. Evolution des populations de bactéries et de meiofaune. *Helgol. Wiss. Meeresunters.* **30**: 163-177.

Characklis, W. G. 1981. Microbial fouling: A process analysis. Pp. 251-291. **In**: D. A. Sommerscales and E. T. Knudsen [eds.], *Fouling of Heat Transfer Equipment*. Washington, D.C.: Hemisphere Publishing Corp.

Christiansen, F. B. and T. M. Fenchel. 1977. *Theories of Populations in Biological Communities.* Berlin: Springer-Verlag. 144 pp.

Conway, H. L. and P. J. Harrison. 1977. Marine diatoms grown in chemostats under silicate or ammonium limitation. IV. Transient responses of *Chaetoceros debilis, Skeletonema costatum* and *Thalassiosira gravida* to a single addition of the limiting nutrient. *Mar. Biol.* **43**: 33-43.

Cooper, D. C. and B. J. Copeland. 1973. Responses of continuous-series estuarine microecosystems to point-source input variations. *Ecol. Monogr.* **43**: 213-236.

Denman, K., W. Calder, C. Davis, S. Demers, M. Estrada, M. Lewis, D. Smith, P. Wangersky, C. Yentsch, and A. Zotin. 1985. Technological developments to implement theory into biological oceanography. Pp. 254-258. **In**: R. E. Ulanowicz and T. Platt [eds.], *Ecosystem Theory for Biological Oceanography. Can. Bull. Fish Aquat. Sci.* **213**.

Droop, M. R. 1976. The chemostat in mariculture. Pp. 71-93. **In**: *10th Europ. Symp. Marine Biol.,* Vol. 1. Wetteren, Belgium: Universa Press.

Droop, M. R. and J. M. Scott. 1978. Steady state energetics of a planktonic herbivore. *J. Mar. Biol. Ass. U.K.* **58**: 749-772.

Dugdale, R. C. 1967. Nutrient limitation in the sea: dynamics, identification, and significance. *Limnol. Oceanogr.* **12**: 685-695.

Egan, B. 1987. Marine microbial adhesion and its consequences. Pp. 220-238. **In**: M. A. Sleigh [ed.], *Microbes in the Sea*. Chichester: Ellis Horwood, John Wiley.

Eppley, R. W. and J. L. Coatsworth. 1968. Nitrate and nitrite uptake by *Ditylum brightwelli*. Kinetics and mechanisms. *J. Phycol.* **4**: 151-156.

Eppley, R. W., J. H. Sharp, E. H. Renger, M. J. Perry, and W. G. Harrison. 1977. Nitrogen assimilation by phytoplankton and other microorganisms in the surface waters of the central North Pacific Ocean. *Mar. Biol.* **39**: 111-120.

Estrada, M., M. Alcaraz, and C. Marrasé. 1987. Effects of turbulence on the composition of phytoplankton assemblages in marine microcosms. *Mar. Ecol. Prog. Ser.* **38**: 267-281.

Fenchel, T. 1977. The significance of bactivorous Protozoa in the microbial community of detrital particles. Pp. 529-544. **In**: J. Cairns [ed.], *Aquatic Microbial Communities*. New York: Garland Publ.

Fenchel, T. 1980. Suspension feeding in ciliated protozoa: feeding rates and their ecological significance. *Microb. Ecol.* **6**: 13-25.

Fera, P., M. A. Siebel, W. G. Characklis, and D. Prieur. 1989. Seasonal variations in bacterial colonisation of stainless steel, aluminium and polycarbonate surfaces in a seawater flow system. *Biofouling* **1**: 251-261.

Giesy, J. P., Jr. [ed.]. 1980. *Microcosms in Ecological Research*. DOE Symposium Series 52, CONF-781101. Springfield, VA: National Technical Information Service. 1110 pp.

Goldman, J. C. 1984. Oceanic nutrient cycles. Pp. 137-170. **In**: M. J. R. Fasham [ed.], *Flows of Energy and Materials in Marine Ecosystems. Theory and Practice.* New York: Plenum Press.

Goldman, J. C., J. J. McCarthy, and D. G. Peavey. 1979. Growth rate influence on the chemical composition of phytoplankton in oceanic waters. *Nature* **279**: 210-215.

Grice, G. D. and M. R. Reeve [eds.]. 1982. *Marine Mesocosms. Biological and Chemical Research in Experimental Ecosystems.* New York: Springer-Verlag. 430 pp.

Gustafsson, K. and L. Gustafsson. 1985. A microcalorimetric perfusion vessel used for measurement of total activity in sediment samples. *J. Microbiol. Methods* **4**: 103-112.

Gustafsson, L. 1987. Microcalorimetry as a tool in microbiology and microbial ecology. Pp. 167-181. **In**: M. A. Sleigh [ed.], *Microbes in the Sea*. Chichester: Ellis Horwood, John Wiley.

Hamilton, W. A. 1987. Biofilms: microbial interactions and metabolic activities. *Sympos. Soc. Gen. Microbiol.* **41**: 361-385.

Hanson, R. B. and K. R. Tenore. 1981. Microbial metabolism and incorporation by the polychaete *Capltella capitata* of aerobically and anaerobically decomposed detritus. *Mar. Ecol. Prog. Ser.* **6**: 299-307.

Hargrave, B. T. and G. A. Phillips. 1981. Annual *in situ* carbon dioxide and oxygen flux across a subtidal marine sediment. *Estuarine Coastal Shelf Sci.* **12**: 725-737.

Henriksen, K., J. I. Hansen, and T. H. Blackburn. 1980. The influence of benthic infauna on exchange of inorganic nitrogen between sediment and water. *Ophelia* Suppl. **1**: 249-256.

Hochachka, P. W. and G. N. Somero. 1973. *Strategies of Biochemical Adaptation*. Philadelphia: W. B. Saunders. 358 pp.

Huq, A., P. A. West, E. B. Small, M. I. Huq, and R. R. Colwell. 1984. Influence of water temperature, salinity, and pH on survival and growth of toxigenic *Vibrio cholerae* serovar 0.1 associated with live copepods in laboratory microcosms. *Appl. Environ. Microbiol.* **48**: 420-424.

Jannasch, H. W. 1967. Enrichments of aquatic bacteria in continuous culture. *Arch. Mikrobiol.* **59**: 165-173.

Jannasch, H. W. 1974. Steady state and the chemostat in ecology. *Limnol. Oceanogr.* **19**: 716-720.

Kristensen, E. and T. H. Blackburn. 1987. The fate of organic carbon and nitrogen in experimental marine sediment systems: Influence of bioturbation and anoxia. *J. Mar. Res.* **45**: 231-257.

Lamprecht, I. and A. I. Zotin [eds.]. 1978. *Thermodynamics of Biological Processes*. Berlin and New York: Walter de Gruyter. 428 pp.

Lasserre, P. 1976. Metabolic activities of benthic microfauna and meiofauna: recent advances and review of suitable methods of analysis. Pp. 95-142. **In**: I. McCave [ed.], *The Benthic Boundary Layer*. New York: Plenum Press.

Lasserre, P. 1980. Energetic role of meiofauna and epifaunal deposit-feeders in increasing level of microbial activity in estuarine ecosystems, at the water-sediment interface. Pp. 309-318. **In**: *Biogéochimie de la Matière Organique à l'interface Eau-Sédiment Marin*. Paris: Editions du CNRS.

Lasserre, P. 1984. The measurement of the enthalpy of metabolism in marine organisms. Pp. 247-269. **In**: M. J. R. Fasham [ed.], *Flows of Energy and Materials in Marine Ecosystems*. New York: Plenum Press.

Lasserre, P. and T. Tournié. 1984. Use of microcalorimetry for the characterization of marine metabolic activity at the water-sediment interface. *J. Exp. Mar. Biol. Ecol.* **74**: 123-139.

Lasserre, P., T. Tournié, M. Bianchi, and S. Chamroux. 1986. Heat production of microorganisms in eutrophied estuarine systems - An experimental study. Pp. 161-174. **In**: P. Lasserre and J. M. Martin [eds.], *Biogeochemical Processes at the Land-Sea Boundary*. Amsterdam: Elsevier.

Ljungholm, K., B. Noren, R. Sköld, and I. Wadsö. 1979. Use of microcalorimetry for the characterization of microbial activity in soil. *Oikos* **33**: 15-23.

Lovitt, R. W. and J. W. T. Wimpenny. 1981. The gradostat, a bidirectional compound chemostat, and its application in microbiological research. *J. Gen. Microbiol.* **127**: 261-268.

MacFarlane, G. T., M. A. Russ, S. M. Keith, and R. A. Herbert. 1984. Simulation of microbial processes in estuarine sediments using gel-stabilized systems. *J. Gen. Microbiol.* **130**: 2927-2933.

Maestrini, S. Y. and D. J. Bonin. 1981. Competition among phytoplankton based on inorganic macronutrients. Pp. 264-278. **In**: T. Platt [ed.], *Physiological Bases of Phytoplankton Ecology. Can. Bull. Fish. Aquat. Sci.* 210.

Margalef, R. 1963. Modelos simplificados del ambiente marino para el estudio de la sucesion y distribucion del fitoplancton y del indicator de sus pigmentos. *Investigacion pesq.* **23**: 11-52.

Margalef, R. 1967. Laboratory analogues of estuarine plankton systems. Pp. 515-524. **In**: G. H. Lauff [ed.], *Estuaries*. Baltimore, Md.: Horn-Shafer.

Margalef, R. 1978. Life-forms of phytoplankton as survival alternatives in an unstable environment. *Oceanol. Acta* **1**: 493-509.

McCarthy, J. J. 1981. The kinetics of nutrient utilization. Pp. 211-233. **In**: T. Platt [ed.], *Physiological Bases of Phytoplankton Ecology. Can. Bull. Fish. Aquat. Sci.* 210.

McCarthy, J. J. and J. C. Goldman. 1979. Nitrogenous nutrition of marine phytoplankton in nutrient-depleted waters. *Science* **203**: 670-672.

McIntyre, A. D., A. L. S. Munro, and J. H. Steele. 1970. Energy flow in a sand ecosystem. Pp. 19-31. **In**: J. H. Steele [ed.], *Marine Food Chains*. Berkeley: Univ. California Press.

Monod, J. 1950. La technique de culture continue; théorie et applications. *Ann. Inst. Pasteur* **79**: 390-410.

Nixon, S. W. 1969. A synthetic microcosm. *Limnol. Oceanogr.* **14**: 142-145.

Nixon, S. W., D. Alonso, M. E. Q. Pilson, and B. A. Buckley. Turbulent mixing in aquatic microcosms. Pp. 818-849. **In**: J. P. Giesy, Jr. [ed.], *Microcosms in Ecological Research*. DOE Symposium Series 52, CONF-781101. Springfield, VA: National Technical Information Service.

Novick, A. and L. Sziland. 1950. Experiments with the chemostat on spontaneous mutations of bacteria. *Proc. Nat. Acad. Sci. U.S.* **36**: 708-719.

de Noyelles, F. and W. J. O'Brien. 1974. The *in situ* chemostat - a self-contained continuous culturing and water sampling system. *Limnol. Oceanogr.* **19**: 326-331.

Odum, H. T., W. L. Siler, R. J. Beyers, and N. Armstrong. 1963. Experiments with engineering of marine ecosystems. *Contrib. Mar. Sci.* **9**: 373-403.

Paffenhöfer, J. 1976. Feeding, growth, and food conversion of the marine planktonic copepod *Calanus helgolandicus*. *Limnol. Oceanogr.* **21**: 39-50.

Painting, S. J., M. I. Lucas, and D. G. Muir. 1989. Fluctuations in heterotrophic bacterial community structure, activity and production in response to development and decay of phytoplankton in a microcosm. *Mar. Ecol. Prog. Ser.* **53**: 129-141.

Pamatmat, M. M. 1965. A continuous-flow apparatus for measuring metabolism of benthic communities. *Limnol. Oceanogr.* **10**: 486-489.

Pamatmat, M. M. 1982. Heat production by sediment: ecological significance. *Science* **215**: 395-397.

Pamatmat, M. M. 1984. Measuring the metabolism of the benthic ecosystem. Pp. 223-246. **In**: M. J. R. Fasham [ed.], *Flows of Energy and Materials in Marine Ecosystems.* New York: Plenum Press.

Pamatmat, M. M., G. Graf, W. Bengtsson, and C. S. Novak. 1981. Heat production, ATP concentration and electron transport activity of marine sediments. *Mar. Ecol. Prog. Ser.* **4**: 234-238.

Parsons, T. R. 1982. The future of controlled ecosystem enclosure experiments. Pp. 411-418. **In**: G. D. Grice and M. R. Reeve [eds.], *Marine Mesocosms. Biological and Chemical Research in Experimental Ecosystems.* New York: Springer-Verlag.

Parsons, T. R., P. J. Harrison, and R. Waters. 1978. An experimental simulation of changes in diatom and flagellate blooms. *J. Exp. Mar. Biol. Ecol.* **32**: 285-294.

Parsons, T. R. and C. A. Bawden. 1979. A controlled ecosystem for the study of the food requirements of amphipod populations. *Estuarine Coastal Shelf Sci.* **8**: 547-553.

Pederson, K. 1982. Method for studying microbial biofilms in flowing water systems. *Appl. Environ. Microbiol.* **43**: 6-13.

Pilson, M. E. and S. W. Nixon. 1980. Marine microcosms in ecological research. Pp. 724-741. **In**: J. P. Giesy, Jr. [ed.], *Microcosms in Ecological Research.* DOE Symposium Series 52, CONF-781101. Springfield, VA: National Technical Information Service.

Platt, T. 1981. *Physiological Bases of Phytoplankton Ecology. Can. Bull. Fish. Aquat. Sci.* **210**. 346 pp.

Platt, T., K. H. Mann, and R. E. Ulanowicz. 1981. *Mathematical Models in Biological Oceanography.* Paris: The Unesco Press. 157 pp.

Prigogine, I. and J. M. Wiame. 1946. Biologie et thermodynamique des phénomènes irréversibles. *Experientia* **2**: 451-453.

Regnault, M. 1986. Production d'NH_4^+ par la crevette *Crangon crangon* L. dans deux écosystèmes côtiers. Approche expérimentale et étude de l'influence du sédiment sur le taux d'excrétion. *J. Exp. Mar. Biol. Ecol.* **100**: 113-126.

Regnault, M., R. Boucher-Rodoni, G. Boucher, and P. Lasserre. 1988. Effects of macrofauna excretion and turbulence on inorganic nitrogenous exchanges at the water-sediment interface. Experimental approach in microcosms. *Cah. Biol. Mar.* **29**: 427-444.

Revsbech, N. P. 1983. *In situ* measurement of oxygen profiles of sediments by use of oxygen microelectrodes. Pp. 265-273. **In**: E. Gnaiger and H. Forstner [eds.], *Polarographic Oxygen Sensors. Aquatic and Physiological Applications*. Berlin & New York: Springer-Verlag.

Rice, D. L. and R. B. Hanson. 1984. A kinetic model for detritus nitrogen: role of the associated bacteria in nitrogen accumulation. *Bull. Mar. Sci.* **35**: 326-340.

Ringelberg, J. 1977. Properties of an aquatic micro-ecosystem. *Helgol. Wiss. Meeresunters.* **30**: 134-143.

Schneider, E. D. 1988. Thermodynamics, ecological succession, and natural selection: a common thread. Pp. 107-138. **In**: B. H. Weber, D. J. Depew, and J. D. Smith [eds.], *Entropy, Information and Evolution.* Cambridge, Mass.: MIT Press.

Scott, J. M. 1980. Effect of growth rate of the food alga on the growth/ingestion efficiency of a marine herbivore. *J. Mar. Biol. Ass. U.K.* **60**: 681-702.

Scott, J. M.. 1985. The feeding rates and efficiencies of a marine ciliate, *Strombidium* sp., grown under chemostat steady-state conditions. *J. Exp. Mar. Biol. Ecol.* **90**: 81-95.

Steele, J. H. 1979. The uses of experimental ecosystems. *Phil. Trans. R. Soc. Lond., B.* **286**: 583-595.

Stephenson, W. and R. B. Searles. 1960. Experimental studies on the ecology of intertidal environments at Heron Island. I. Exclusion of fish from beach rock. *Aust. J. Mar. Freshwat. Res.* **11**: 241-267.

Taub, F. B. 1969. A biological model of a freshwater community: a gnotobiotic ecosystem. *Limnol. Oceanogr.* **14**: 136-142.

Taub, F. B. and M. E. Crow. 1980. Synthesizing aquatic microcosms. Pp. 69-104. **In**: J. P. Giesy, Jr. [ed.], *Microcosms in Ecological Research.* DOE Symposium Series 52, CONF-781101. Springfield, VA: National Technical Information Service.

Tenore, K. R. 1987. Nitrogen in benthic food chains. Pp. 191-206. **In**: T. H. Blackburn and J. Sorensen [eds.], *Nitrogen Cycling in Coastal Marine Environments.* New York: Wiley.

Tenore, K. R., R. B. Hanson, J. McClain, A. E. MacCubbin, and R. E. Hobson. 1984. Changes in composition and nutritional value to a benthic deposit feeder of decomposing detritus pools. *Bull. Mar. Sci.* **35**: 299-311.

Tournié, T. and P. Lasserre. 1984. Microcalorimetry characterization of seasonal metabolic trends in marine microcosms. *J. Exp. Mar. Biol. Ecol.* **74**: 111-121.

Valiella, I., J. Wilson, R. Buchsbaum, C. Rietsma, D. Bryant, K. Foreman, and J. Teal. 1984. Importance of chemical composition of salt marsh litter on decay rates and feeding by detritivores. *Bull. Mar. Sci.* **35**: 261-269.

Veldkamp, H. and H. W. Jannasch. 1972. Mixed culture studies with the chemostat. *J. Appl. Chem. Biotechnol.* **22**: 105-123.

Vernberg, F. J. and W. B. Vernberg. 1981. *Functional Adaptations of Marine Organisms.* New York: Academic Press. 347 pp.

Whittaker, R. H. 1961. Experiments with radiophosphorus tracer in aquarium microcosms. *Ecol. Monogr.* **31**: 157-187.

Wimpenny, J. W. T. 1981. Spatial order in microbial ecosystems. *Biol. Rev.* **56**: 295-342.

Wimpenny, J. W. T. and A. Peters. 1987. Ecology on the microscale. Pp. 59-82. **In**: M. A. Sleigh [ed.], *Microbes in the Sea*. Chichester: Ellis Horwood, John Wiley.

Wimpenny, J. W. T., J. P. Coombs, R. W. Lovitt, and A. Whittaker. 1981. A gel-stabilized model ecosystem for the investigation of microbial growth in spatially ordered solute gradients. *J. Gen. Microbiol.* **127**: 277-287.

Yentsch, C. M., P. K. Horan, K. Muirhead, Q. Dortch, E. Haugen, L. Legendre, L. S. Murphy, M. J. Perry, D. A. Phinney, S. A. Pomponi, R. W. Spinrad, M. Wood, C. S. Yentsch, and B. J. Zhahuranec. 1983. Flow cytometry and cell sorting: a technique for analysis and sorting of aquatic particles. *Limnol. Oceanogr.* **28**: 1275-1280.

Zeitzschel, B. 1981. Field experiments on benthic ecosystems. Pp. 607-625. **In**: A. Longhurst [ed.], *Analysis of Marine Ecosystems.* New York: Academic Press.

Zotin, A. I. 1985. Thermodynamics and the growth of organisms in ecosystems. Pp. 27-37. **In**: R. E. Ulanowicz and T. Platt [eds.], *Ecosystem Theory for Biological Oceanography*. *Can. Bull. Fish Aquat. Sci.* **213**.

4. PELAGIC MESOCOSMS:
I. FOOD CHAIN ANALYSIS

Masayuki Takahashi

Abstract

Mesocosm experiments conducted for the study of pelagic plankton ecosystems have been reviewed. The capability of maintaining a given ecosystem in a mesocosm under near-natural conditions and the reproducibility of biological events in multiple mesocosms were evaluated. Mesocosm experiments were tentatively grouped as follows: (1) experimental studies using mesocosms under various manipulations; (2) budget estimates of materials and energy in ecosystems; and (3) pollutant effects on the organisms and the ecosystem.

INTRODUCTION

Looking at the history of the development of natural science, it is obvious that field observations were the first approach for most of natural science; this resulted in finding many scientific truths about nature. However, there were some phenomena requiring further evaluation under less complicated situations than nature. Laboratory experiments were conducted in order to supplement the field observations, and these also contributed extensively to the finding of scientific truths about nature (Banse, 1982).

There are some drawbacks to laboratory experiments (Strickland and Terhune, 1961). One of these is that only selected species (i.e. not always dominants but minorities in nature) are experimentally cultured in the laboratory (Parsons, 1988); another is that multispecies systems are almost impossible to maintain in the laboratory except in extremely limited cases. Consequently, laboratory experiments have contributed to the finding of truth mainly on a given single population (i.e. that which can be maintained in the laboratory).

There is still a big question left behind in nature, which is how to synthesize various basic processes demonstrated by laboratory experiments for the understanding of nature (reductionist approach), and even whether it is possible for such synthesizing of the results from a reductionist approach. The importance of a holistic approach has been stressed, particularly for the understanding of ecosystems (Odum, 1984).

Capturing part of a natural ecosystem in a certain size of enclosure (i.e. a mesocosm) eliminates some drawbacks occurring in laboratory experiments. Major advantages of mesocosm experiments are that a multispecies natural community can be maintained for a prolonged period of time and that some biological and abiological environmental factors can be controlled (Gamble and Davies, 1982). Of course some new drawbacks created by mesocosms occur due to setting spatial boundaries, which are variable and usually unknown in natural systems. Spatial boundaries eliminate the lateral watermass movement and some wave action (Steele et al., 1977) and, in turn, tend to supply a new platform for particular groups of organisms such as those which cause fouling. In an ecosystem, furthermore, there is a large range of spatial and temporal scales occupied by different biological components. The distances between naturally occurring communities vary from a few kilometers for phytoplankton to hundreds of kilometers for fish, while zooplankton are intermediate. In addition, since the understanding of population interactions and fluctuations requires a minimum of one and up to several generation times of the organisms, temporal studies may require from days to weeks for phytoplankton, months for zooplankton, and years for fish (Menzel, 1980). With these changes in the mesocosm, an enclosed ecosystem deviates with time compared to the surrounding natural ecosystem. It is very important to recognize that the natural environment cannot be exactly duplicated in mesocosms. Nevertheless, mesocosms can be extremely useful tools when employed to answer specific questions that are not easily addressed, either by simple field observations or laboratory experiments.

It must therefore be kept in mind in using mesocosms that the target phenomenon which is related to the question to be answered must be maintained in the mesocosm, and furthermore that the phenomenon must not be affected by changes due to the mesocosm structure. Types and sizes of mesocosms and the time period of experiments will be variable depending upon the question(s). Great effort has been made to create near-natural conditions in mesocosms, or at least to maintain natural multispecies communities and multitrophic levels of ecosystems in mesocosms. Scientific studies resulting from mesocosm experiments for pelagic plankton ecosystems fall into three groups. There are: (1) experimental studies using mesocosms under various manipulations; (2) budget estimates of materials and energy in ecosystems; and (3) pollutant effects on the organisms and the ecosystem.

RESULTS AND DISCUSSION

Reproducibility of Pelagic Plankton Ecosystems in Mesocosms

One great advantage of using a mesocosm is its potential capability to capture and maintain a natural ecosystem having multispecies with multitrophic levels, which is partly self-controlled by feedback mechanisms within the mesocosm. However, as mentioned earlier, the captured ecosystem cannot be expected to maintain all the processes occurring in the natural environment, and it can both eliminate and modify some processes as well as adding new processes. These changes are physical, chemical, and biological. For example, there is an elimination of nekton and large zooplankton during the capturing operation because these creatures are too active and have a large requirement for space in which to survive. Recruitment of and recolonization by organisms from surrounding areas are also blocked in mesocosms, unless there is artificial introduction of those particular organisms. The captured ecosystem, therefore, tends to deviate from the surrounding system with time, although some efforts can be made in order to minimize the deviations.

Reproducibility experiments of the spring phytoplankton bloom and of upwelling ecosystems simulated in mesocosms of 125 m^3 "balloon" and 60 m^3 "tube" types can be mentioned as examples (McAllister et al., 1961; Antia et al., 1963; Takahashi et al., 1975). In these experiments, a large volume of water was captured in a certain size of mesocosm made of transparent plastic, and the biological events within the mesocosms were followed over time with the least manipulation in order to simulate near-natural conditions. Biological and some other parameter changes were followed with time, during which multispecies phytoplankton grew rapidly by consuming nutrients and formed a bloom. The growth kinetics and chemical composition of the mixed population were related to the environment with a minimum of ambiguity during the experimental periods of 30 to 90 days. Enhancement of herbivorous zooplankton due to the massive development of phytoplankton was also reproduced in the mesocosms.

Organisms at the higher trophic levels often require special attention, including separate introduction, in order to be maintained in a mesocosm. Fish are extremely sensitive in their requirements for survival, and maintaining their growth in mesocosms is difficult because of their specific food requirements. Young salmonids (Koeller and Parsons, 1977; Sonntag and Parsons, 1979) and juvenile herring (Steele and Gamble, 1980; Houde and Berkeley, 1982) have been tested. Many mesocosm experiments tended to develop a ctenophore population as the dominant top carnivore (Parsons et al., 1977a; Harris et al., 1980). There is a general trend over a period of time for a steady decrease in species diversity in a mesocosm (Parsons et al., 1977b). At present, it has been impossible to devise a water column which accommodates survival of strongly migratory species, such as euphausiids, and attempts to introduce them into even the largest plastic columns of 1,300 m^3 have resulted in their rapid demise (Parsons, 1981).

By using multiple mesocosms, replication of biological events between mesocosms has been reasonable in that the patterns of temporal changes remained similar for a few months. However, absolute values may vary to some degree and deviations become larger with time, even if the water column was captured simultaneously in all the mesocosms used (Fig. 1; cf. Takahashi et al., 1975). The temporal deviation seems to become progressively greater at higher trophic levels, and replication has become more difficult as mesocosms have increased in size.

Experiments have confirmed that natural mixed populations having multitrophic levels can be maintained in mesocosms, although some organisms require special care. Major processes of natural biological events can also be duplicated in mesocosms. It is a remarkable opportunity to be able to follow the same ecosystem repeatedly; this can give us a good time-series of ecosystem data because of less effect by lateral watermass movement and other disturbances occurring in the sea (Davis et al., 1980). Consequently, it has become obvious that mesocosms supply us with experimental ecosystem data under near-natural conditions. However, the duration for mesocosm experiments is rather short. Experiments of a few weeks to a few months accommodate the generation time of lower trophic level organisms but are not long enough for generations of organisms at higher trophic levels (Menzel, 1980).

Mesocosm Experiments under Various Manipulations of the Ecosystem

One of the most attractive points of mesocosm experiments is that, given a target process in an ecosystem, it can be enhanced or eliminated by artificial manipulation, and the importance of the target process in the ecosystem can then be evaluated both qualitatively and quantitatively. The following are several examples of experiments conducted along this line.

Experimental Control of Dominant Groups of Organisms in Pelagic Ecosystems

One of the basic general questions in this type of research is to find possible mechanisms involved in the determination of a given population structure in a pelagic ecosystem. Major changes in dominant phytoplankton are controlled by short-term variations in the physical and chemical environment or by grazing pressure exerted by herbivores. The sequence of events from a diatom to a flagellate bloom was simulated in a large mesocosm

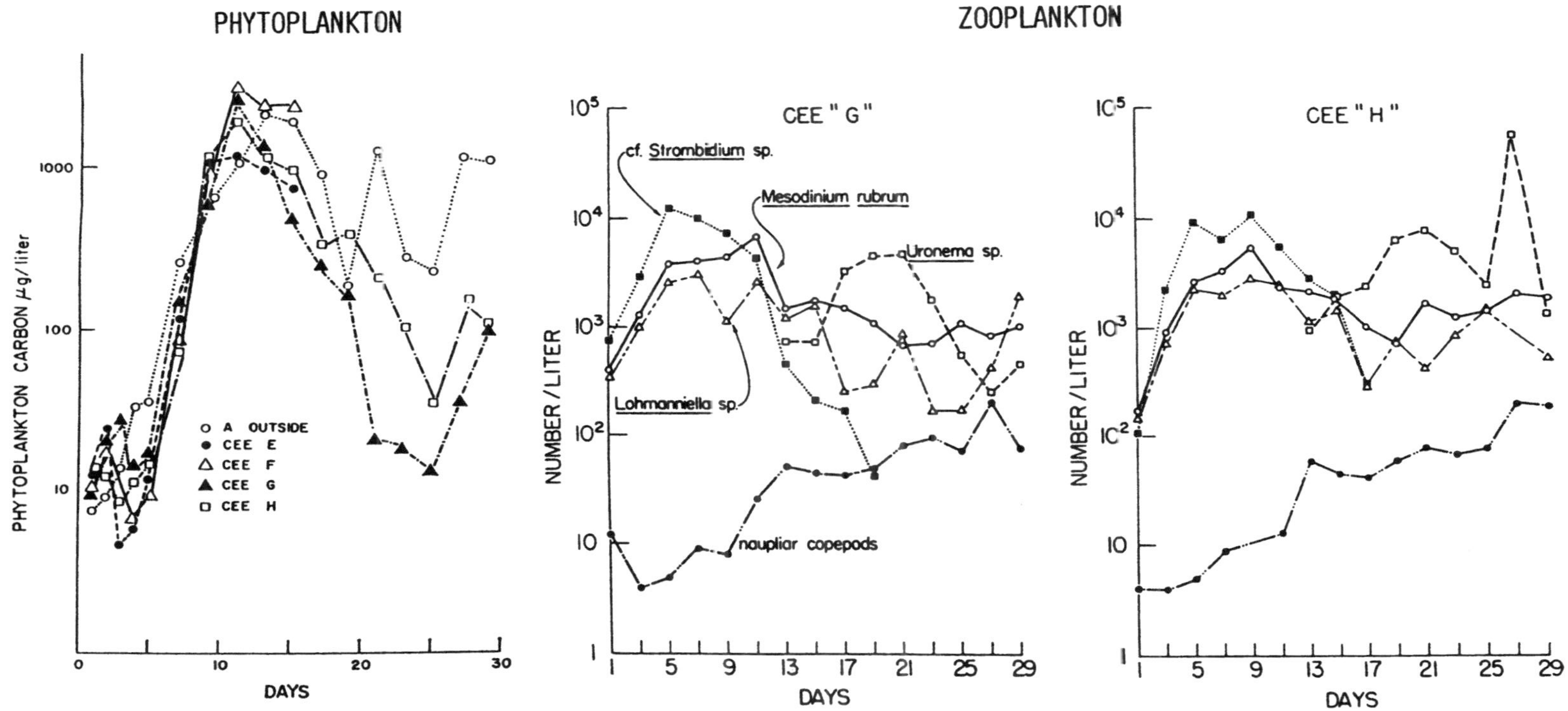

Figure 1. Temporal changes in total phytoplankton and individual groups of dominant zooplankton in identically treated multiple mesocosms (60 m^3, cylinder type) during one month. CEE, Controlled ecosystem enclosure. After Takahashi et al. (1975).

of 1,300 m^3 under controlled light intensity and nutrients over 112 days (Parsons et al., 1978). The results indicated that growth of large diatoms could be manipulated to occur under specific conditions of light intensity and nutrient concentrations, whereas growth of the small flagellate occurred under a wide range of environmental changes (Fig. 2). Diatom dominance appeared to be due to the efficient and rapid multiplication of cells under specific favorable conditions compared with the small flagellates. These results broadly explained the general geographical distributions and temporal occurrences of these two groups of algae in pelagic ecosystems. The specific conditions favorable for diatoms occur in upwelling areas and during spring in temperate waters. Under these conditions, large, and at times almost unialgal, diatom communities develop. In other areas where those specific conditions are not satisfied, communities with a high species diversity of small flagellates and monads become dominant.

Grice et al. (1980) created diatom- and flagellate-dominated ecosystems experimentally in two separate 1,300 m^3 mesocosms simultaneously over one month. This was done by introducing periodic nutrient additions, light shielding, and turbulence (bubbling for the diatom ecosystem, and by nutrient additions without silicate for the flagellate ecosystem).

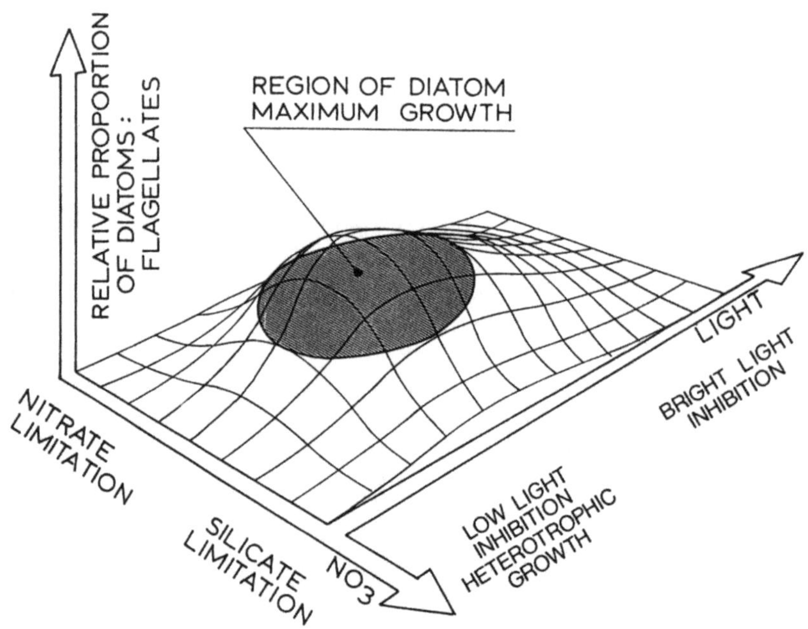

Figure 2. Hypothetical physico-chemical conditions that favor diatoms and flagellates. From Parsons et al. (1978).

These two examples given above were both conducted under insignificant grazing pressure by herbivorous animals on the phytoplankton population. In the experiment done by Grice et al. (1980), a large number of ctenophores developed and effectively eliminated herbivorous copepods from the mesocosms. It is highly probable that such a large mesocosm may not be required to control the population structure of phytoplankton by regulating the physico-chemical conditions, but that a smaller size is often adequate.

If the activity of herbivorous animals is high, the population structure of phytoplankton will also be affected significantly by those animals. Steele and Gamble (1980) created two typical phytoplankton populations under different levels of predator pressure on herbivorous animals: high predator numbers caused low herbivore populations which resulted in large diatom populations; low predator numbers led to more herbivores and resulted in small flagellate populations due to active removal of diatoms from the system by herbivores.

These experiments revealed that phytoplankton populations can be controlled either from the physico-chemical environmental conditions (bottom-up control) or by the grazing pressure of herbivorous animals (top-down control). Similarly, two levels of control can be anticipated for organisms at other trophic levels.

Other controls of phytoplankton populations have been approached using laboratory cultures in a 70 m^3 land-based mesocosm (Strickland et al., 1969). A centric diatom (*Ditylum brightwellii*) and a dinoflagellate (*Cachonia niei*) were grown individually in mesocosms, and a dinoflagellate (*Gonyaulax polyedra*) and a haptophyte (*Phaeocystis* sp.) were cultured together. Based upon these experiments, favorable growth conditions for each phytoplankton population could be experimentally evaluated under near-natural water column conditions.

Experimental Control of Growth Phasing of Populations in Pelagic Ecosystems

As every population at different trophic levels in the ecosystem is connected by the prey-predator relationship, temporal changes in populations are required to be phased to each other at least to some extent. Otherwise the predator cannot maintain the population size corresponding to the prey density and, furthermore, the prey may not be able to maintain its population because of insufficient recovery of nutrients, unless there is a new source of nutrients other than from recycling in the ecosystem. To support a healthy ecosystem driven by efficient transfers of materials and energy between trophic levels, organisms in the prey-predator relationships are required to be phased in their temporal changes.

The first example on phasing involved control of the extent of a phytoplankton bloom by reducing available solar radiation in the ecosystem through the addition of mine tailings into mesocosms (Parsons et al., 1986). The three 60 m^3 mesocosms were treated with 0, 39, and 297 ppm of mine tailings which had very low biological activity except for the fact that they caused severe light attenuation in the mesocosms. The 1% light level was reduced to less than 2 m in the mesocosms with mine tailings compared to about 10 m in the mesocosms with no tailings. This reduction in light attenuation delayed the phytoplankton bloom, which resulted in a greater biomass of zooplankton (Fig. 3). These results were interpreted as showing that increased turbidity caused phytoplankton production to be slowed down, and that consequently the zooplankton were able to take greater advantage of the phytoplankton standing stock.

Differences in the ecosystem production observed in different biomes can often be attributed to the phasing conditions of the dominant populations. Examples include the low production at higher trophic levels in a spring biome compared with that of upwelling (Sonntag and Parsons, 1979); high zooplankton biomass with low phytoplankton biomass in the North Pacific, compared to low zooplankton biomass with high phytoplankton biomass in the North Atlantic biome (Parsons and Lalli, 1988); and large discrepancies in fisheries production between locations and seasons compared to the magnitude of primary production in the South Atlantic Bight (Menzel, 1980).

Another example is found in the summer decrease of a pennate diatom, *Asterionella formosa*, in lakes (Kudoh, 1988). Natural population change of *A. formosa* was determined by the instantaneous growth and decay rates under given physico-chemical conditions. The latter was assumed mostly due to fungal infection activity, which is highly dependent on temperature and the population density of *A. formosa* in the water column. A hypothesis, that the *in situ* growth rate of the *A. formosa* population becomes negative in summer due to a high loss rate by fungal infection at higher temperatures compared to the instantaneous growth rate, was proposed and tested experimentally using mesocosms in a eutrophic lake. The instantaneous growth rate was enhanced by reducing the mixing depth of the water column using a shallow mesocosm. This was possible because the euphotic zone was approximately 1 m, and the entire water column of 4 m mixed vertically at least once per day by thermal convection and wind forcing under natural conditions. The *A. formosa* population obviously increased its biomass in the shallow mesocosm, but it decreased constantly in the deep mesocosm (Fig. 4). The results suggest that water mixing extending below the euphotic zone diminished the population size of *A. formosa* in the water column, which explained the summer disappearance of the species.

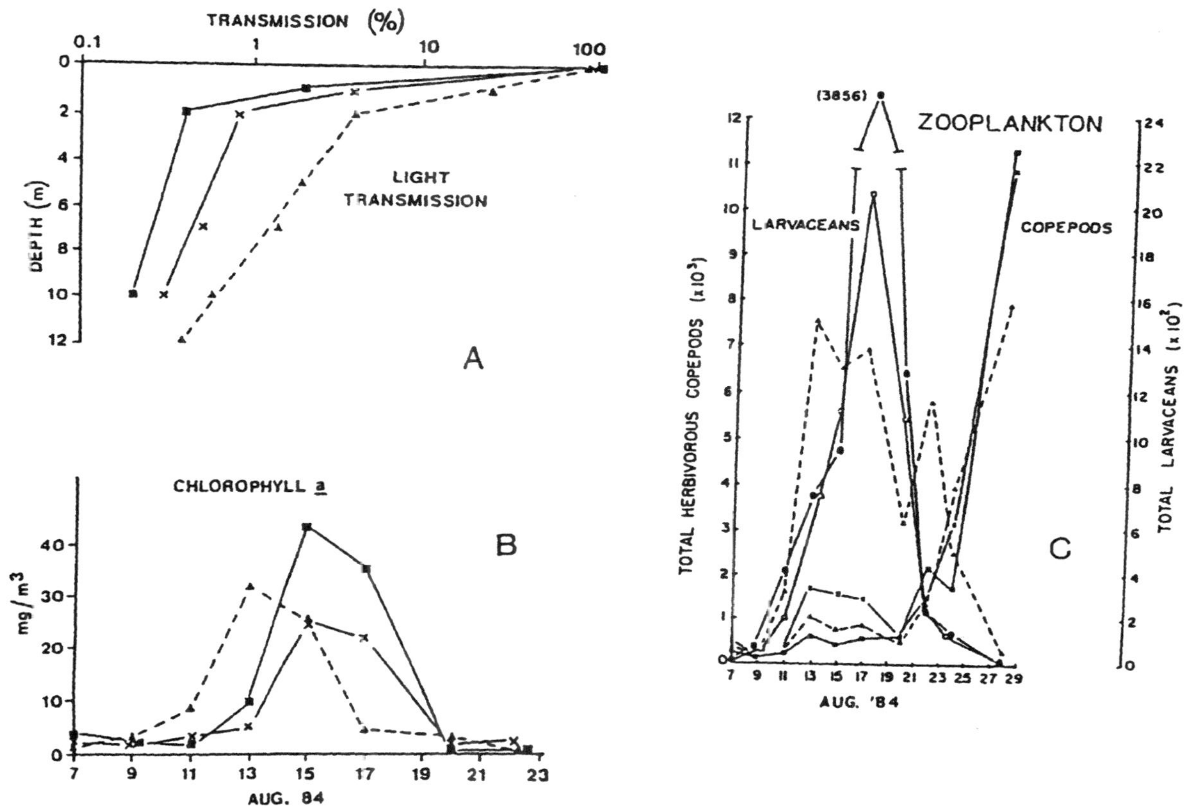

Figure 3. Growth rate control of phytoplankton populations by mine tailings and its effects on the development of herbivorous zooplankton. Dotted line, control; x, 39 ppm tailings; ■, 297 ppm tailings. From Parsons et al. (1986).

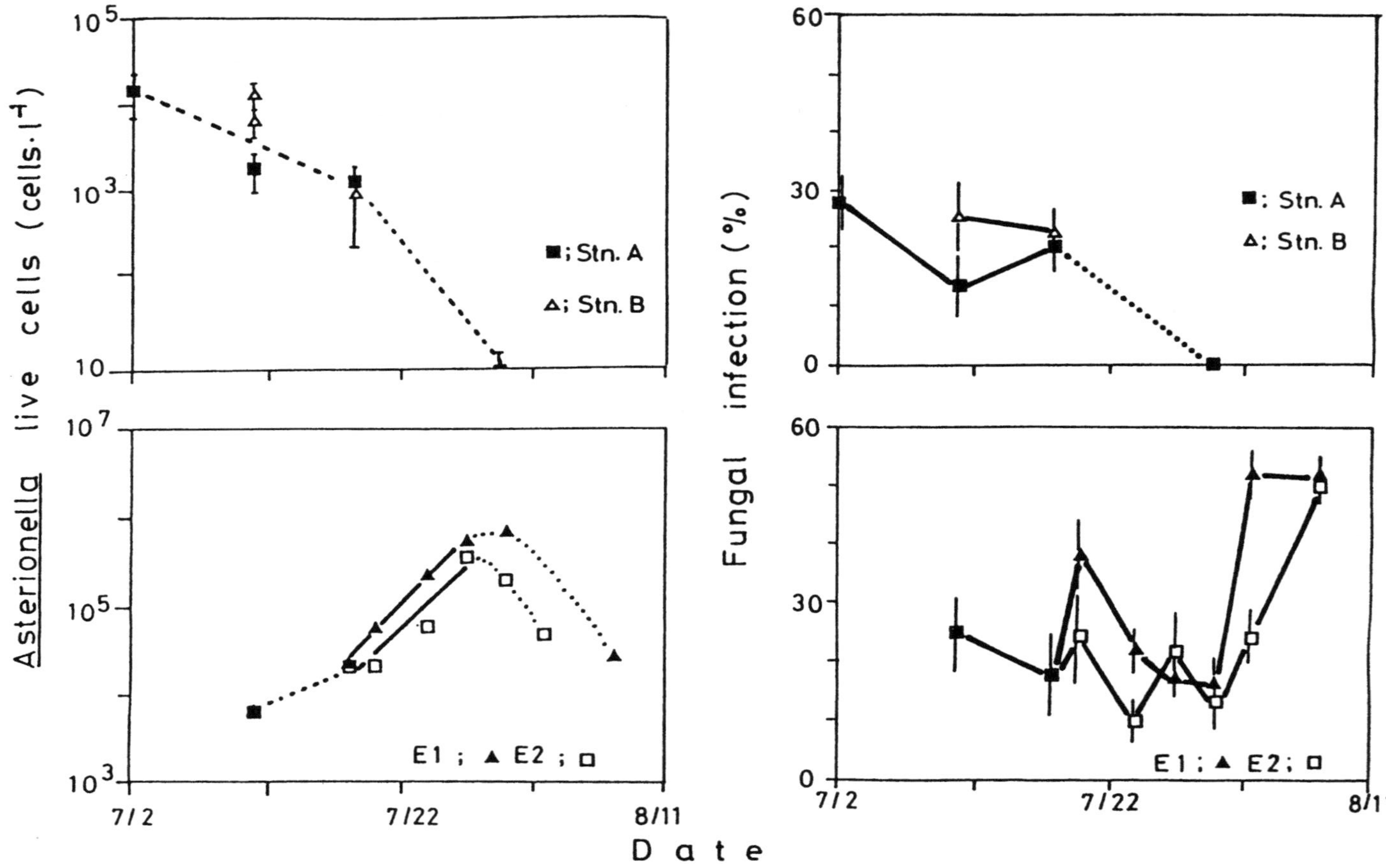

Figure 4. Temporal changes in a *Asterionella formosa* population (left figures) and fungal infection (right figures) in two stations, A and B, of a lake (upper figures) and in shallowed mesocosms, E1 and E2 (lower figures). From Kudoh (1988).

Evaluation of Less Sensitive Biological Events for Given Manipulations

Grice et al. (1980) found apparently synchronous development of a dinoflagellate, *Ceratium furca*, in two mesocosms 50 days after their experiment was started, even though radically different manipulations were imposed and large effects were obvious in diatoms and many of the flagellate community. This suggests that factors other than those which were manipulated were important in the development of the particular species of dinoflagellate bloom.

Less sensitivity of biological events was observed in experiments conducted in a shallow eutrophic lake, in which two of three mesocosms of 5m x 5m x 4m (depth) were covered with a sheet to give 100%, 10% or 1% light transmittance for 60 days (Takahashi et al., 1988). All three mesocosms included bottom sediments. Photosynthetic production decreased drastically in the mesocosm covered by the 1% transmittance sheet, due to a radical decrease of incoming solar radiation. However, the temporal changes of most of the phytoplankton and zooplankton populations followed the same sequences between mesocosms under the three different radiation treatments (Fig. 5). The results suggested that the natural ecosystem has some feedback system which passively or actively makes the ecosystem stable and damps out environmental disturbance, as suggested by Patten and Odum (1981). These results indicated that the experimental mesocosm approach is quite useful in determining the sensitivity of biological events to a given environmental perturbation.

Determination of the Food Chain Pathways of Pelagic Ecosystems

For the analysis of food chains of planktonic ecosystems, stable isotopes of ^{13}C and ^{15}N were used in 2 m^3 mesocosms in a eutrophic freshwater lake (Sakamoto et al., 1988). Two different size groups of phytoplankton, a large cultured diatom, *Diatoma*, and a naturally-occurring community dominated by a small diatom, *Cyclotella*, were labelled by ^{13}C and ^{15}N, and then introduced into a community enclosed in mesocosms. Animals in various taxonomic groups and at various growth stages were sampled from the mesocosm with time, and their ^{13}C and ^{15}N contents were determined. The results clearly showed which herbivores grazed which food items and, furthermore, which herbivorous animals were eaten by which carnivores. The experiment also gives us the time duration of material transfer from one trophic level to the next. This is a fairly direct approach for the determination of a food web system, although it requires a great deal of manpower in order to gather enough samples for the determination of ^{13}C and ^{15}N content in each group.

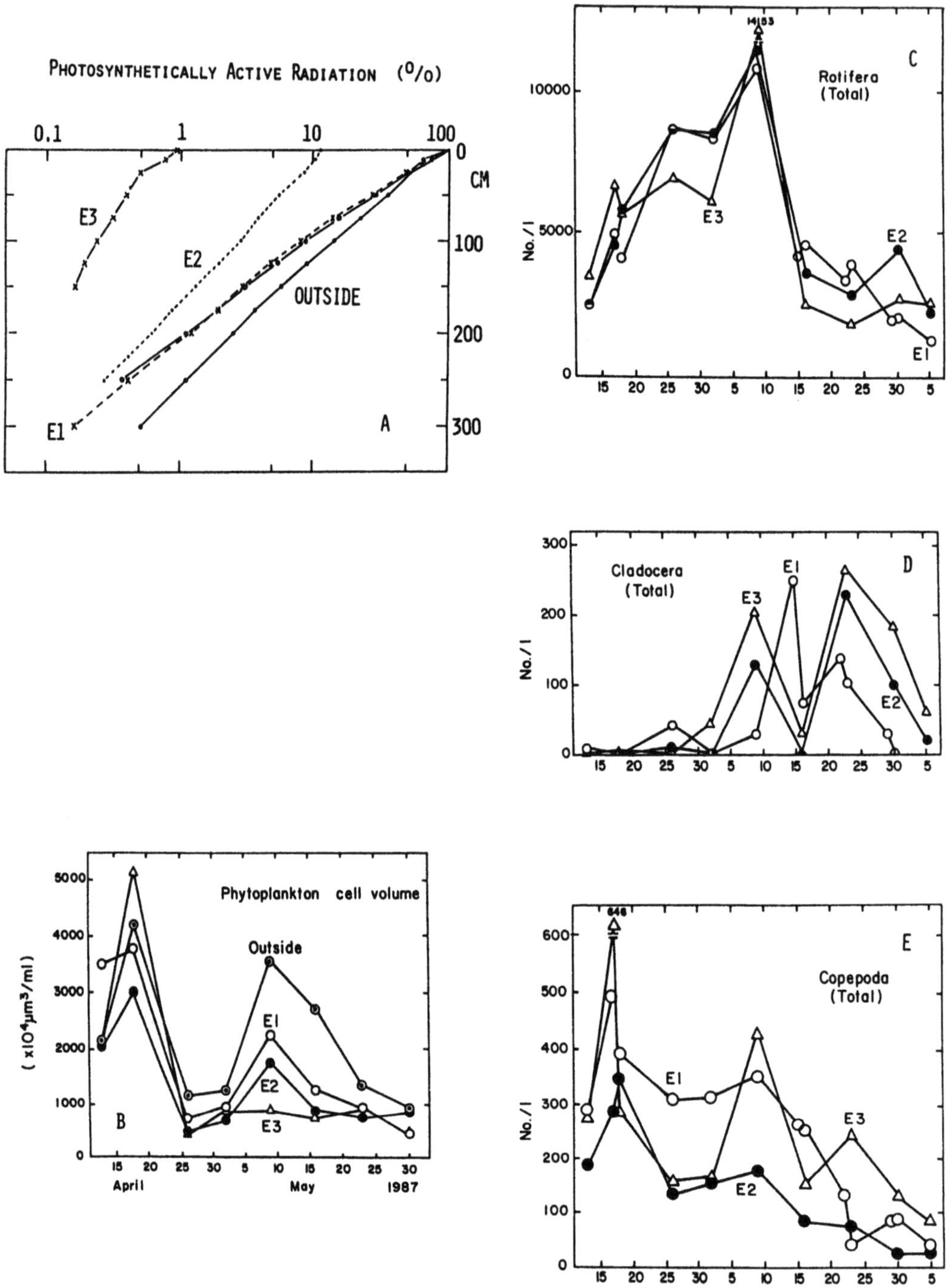

Figure 5. Temporal changes in phytoplankton (B) and three major groups of zooplankton (C, D, and E) in mesocosms covered with no (E1), 10% (E2) and 1% (E3) light transmittance sheet. Fig. 5A shows the light transmittance in the water column relative to the surface irradiance received in the mesocosm with no shading sheet. From Takahashi et al. (1988).

A few experiments were conducted by introducing labelled inorganic nutrients (^{15}N-labelled nitrogen compounds, ^{13}C-carbonates, or several radioisotope-labelled compounds) into mesocosms for the determination of food web structure (cf. Otsuki et al., 1985). Hattori et al. (1980) introduced ^{15}N-labelled ammonium and nitrate into 60 m^3 mesocosms, and determined directly the uptake rates for those compounds and the regeneration rate of ammonium under near-natural conditions.

The other approach for food web analysis is to change the structure and/or the standing stock size of a given food level, and then determine the possible effect(s) on some other level. Grice et al. (1980) have tried to evaluate the effect of quality difference in primary producers on the higher trophic levels. However, the experiment was not successful from that point of view because of a rapid decrease of herbivorous animals due to heavy predation by carnivores. Biomass control of the higher trophic level organisms can be made by removal, using nets towed in the mesocosm or by introducing chemicals (dry ice, pesticides and so on), and/or by addition of a particular group (or groups) of organisms. All these controls can be made more effectively in mesocosms than in the open sea.

Determination of Food Chain and Ecological Efficiencies in Pelagic Ecosystems

Material and energy transfer between trophic levels in the same ecosystem can be followed over time in mesocosms. Parsons et al. (1977a) estimated the transfer efficiency of carbon between trophic levels of dominant plankton from the temporal changes in standing stocks of organisms and their productivities at various trophic levels, using four 65 m^3 mesocosms in a coastal planktonic community over two weeks. They further tested the effects of low level nutrient additions on the productivity and the transfer efficiency between trophic levels. The results showed that instantaneous nutrient addition increased production at all the trophic levels, but decreased the transfer efficiency between each trophic level. Sonntag and Parsons (1979) came to the same conclusion in the experiments using three, even larger, 1,300 m^3 mesocosms in which nutrients were manipulated by bubbling.

Mullin and Evans (1974) determined transfer efficiencies between food levels using a linear food chain consisting of the chain-forming diatom *Skeletonema costatum*, the copepods *Acartia tonsa* and *Paracalanus parvus*, and the ctenophore *Pleurobrachia bachei* in a 70 m^3 land-based tank. The system was maintained over 180 days by making additions of the cultured alga to the tank. The system was not at steady state, but showed clear temporal population changes from *A. tonsa* to *P. parvus*, and then to *P. bachei*.

Figure 6 represents a summary of the transfer efficiencies of planktonic food chains in which ctenophores were the top carnivores. These data were obtained from three different areas, using entirely different approaches and covering a wide range of communities from the natural to an experimental, linear, food chain created in a land-based tank.

Evaluation of Pollutant Effects on the Pelagic Plankton Ecosystem

Environmental perturbations have been made by introducing various organic and inorganic pollutants into experimental pelagic ecosystems. Concentrations of added pollutants can be controlled at proposed levels without the difficulties of infinite dispersion or of a large spatial heterogeneity of distribution, which is extremely difficult to follow in nature. The effect of environmental stress on individual organisms and their recovery in the ecosystem are directly related to the generation time and individual size of the organisms. The larger organisms are generally more sensitive to pollutants and their recovery takes a longer period of time, or the recovery may never occur if they are affected severely. Changes in bacterial activity were detected within hours after pollutants were added, but recovery was rapid within a few days (Azam et al., 1977). Phytoplankton growth was reduced or increased within a few days depending upon the species and the concentrations of pollutant, but it generally returned to control levels within 1 to 2 weeks (Lee et al., 1977; Thomas et al., 1977). The recovered population showed little effect from the existing pollutant, but often the dominant species was changed. Alterations in the zooplankton biomass and species composition were mostly size dependent. Parts of these populations recovered in 30 days or so but, in some cases, recovery was not detected even after 90 days (Grice et al., 1977).

It seems that pollutant stress, whatever its cause, has the universal impact of causing marine plant communities dominated by large diatoms to become dominated by small species, most frequently by flagellates (Menzel, 1980). This has important implications because, as the size of primary producers is reduced, the number of trophic levels is increased; this decreases the total predator yield (Ryther, 1969) and alters the basic structure of the prey-predator community (Greve and Parsons, 1977). Changes created by pollutants in the timing and phasing of biological events also may cause significant effects on food web dynamics (Parsons, 1988).

The other aspect of pollutant effects on the pelagic ecosystem is bioaccumulation of pollutants through the physiological processes of each organism and the prey-predator processes in food web dynamics. Both of these processes can also be evaluated effectively by using mesocosms.

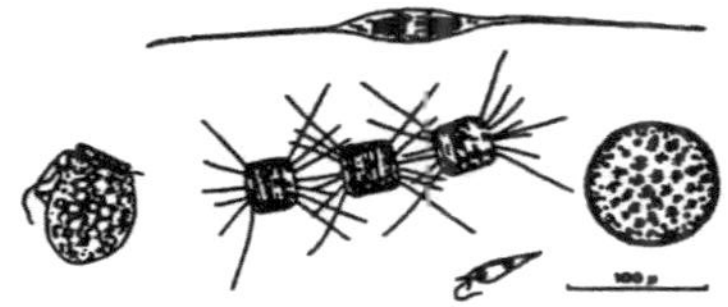

Increased community complexity ↑

Increased artificial manipulation ↓

	PHYTOPLANKTON PRIMARY PRODUCTION (mgC/m^2-day)	HERBIVOROUS ZOOPLANKTON PRODUCTION (% of primary production)	CARNIVOROUS CTENOPHORES PRODUCTION (% of primary production)
PETIPA, PAVLOVA & MIRONOV 1970 (Black Sea, early summer)	854 (measured)	23	3.1 (all carnivores)
PARSONS *et al* 1977 (CEPEX bag, 2 weeks)	109 (unenriched) 306 (enriched)	42 16	5.7 3.5
MULLIN & EVANS 1974 (SIO deep tank, 75 days)	49 (controlled)	>12.5 >10.4	2.6 - 3.2

Figure 6. Production at three major trophic levels determined by three different approaches from field observations (Petipa et al., 1970) to ecosystem experiments (Mullin and Evans, 1974; Parsons et al., 1977). After Mullin (1980).

CONCLUSIONS

It has been confirmed that a given pelagic plankton ecosystem can be maintained under near-natural conditions at least for a few weeks to a few months, although with time the biological events progressively deviate from the outside system and, but with lesser magnitude, between mesocosms. Many previous mesocosm experiments indicate the great usefulness of the mesocosm approach to test various kinds of hypotheses which were delineated from field observations and laboratory experiments. Future mesocosm experiments, therefore, need to be carried out with much closer connections between laboratory and field studies in order to demonstrate their further usefulness in ecosystem understanding.

Some of the disadvantages of the mesocosm approach for the study of pelagic ecosystems lie in the fragile structure of most mesocosms and in the large amount of daily manual care they require. Consequently, past mesocosm experiments, even for pelagic ecosystems, were mostly carried out in near-shore areas protected from strong wind and waves, and having easy access. This has partly limited the kinds of research subjects. The free-floating system designed by Brockmann et al. (1983) represents an advance in design and a necessary progression to the oceanic environment.

LITERATURE CITED

Antia, N. J., C. D. McAllister, T. R. Parsons, K. Stephens, and J. D. H. Strickland. 1963. Further measurements of primary production using a large-volume plastic sphere. *Limnol. Oceanogr.* **8**: 166-183.

Azam, F., F. R. Vaccaro, P. A. Gillespie, E.-I. Moussali, and R. E. Hodson. 1977. Controlled ecosystem pollution experiment: Effect of mercury on enclosed water columns. II. Marine bacterioplankton. *Mar. Sci. Comm.* **3**: 313-329.

Banse, K. 1982. Experimental marine ecosystem enclosures in a historical perspective. Pp. 239-249. **In**: G. D. Grice and M. R. Reeve [eds.], *Marine Mesocosms. Biological and Chemical Research in Experimental Ecosystems.* New York: Springer-Verlag.

Brockmann, U. H., E. Dahl, J. Kuiper, and G. Kattner. 1983. The concept of POSER (Plankton Observation with Simultaneous Enclosures in Rosfjorden). *Mar. Ecol. Prog. Ser.* **14**: 18.

Davis, S. D., J. T. Hollibaugh, D. L. R. Seibert, W. H. Thomas, and P. J. Harrison. 1980. Formation of resting spores by *Leptocylindrus danicus* (Bacillariophyceae) in a controlled experimental ecosystem. *J. Phycol.* **16**: 296-302.

Gamble, J. C. and J. M. Davies. 1982. Application of enclosures to the study of marine pelagic systems. Pp. 25-48. **In**: G. D. Grice and M. R. Reeve [eds.], *Marine Mesocosms. Biological and Chemical Research in Experimental Ecosystems.* New York: Springer-Verlag.

Greve, W. and T. R. Parsons. 1977. Photosynthesis and fish production: The possible effects of climatic change and pollution. *Helgol. Wiss. Meeresunters.* **30**: 666-672.

Grice, G. D., M. R. Reeve, P. Koeller, and D. W. Menzel. 1977. The use of large (1300 m^3), transparent, enclosed sea-surface water columns in the study of stress on plankton ecosystems. *Helgol. Wiss. Meeresunters.* **30**: 118-133.

Grice, G. D., R. P. Harris, M. R. Reeve, J. F. Heinbockel, and C. D. Davis. 1980. Large-scale enclosed water-column ecosystems. An overview of Foodweb I. The final CEPEX experiment. *J. Mar. Biol. Ass. U.K.* **60**: 401-414.

Harris, R. P., M. R. Reeve, G. D. Grice, G. F. Evans, V. R. Gibson, J. R. Beers, and B. K. Sullivan. 1980. Trophic interactions and production processes in natural zooplankton communities in enclosed water columns. Pp. 353-387. **In**: G. D. Grice and M. R. Reeve [eds.], *Marine Mesocosms. Biological and Chemical Research in Experimental Ecosystems.* New York: Springer-Verlag.

Hattori, A., I. Koike, M. Ohtsu, J. J. Goering, and D. Boisseau. 1980. Uptake and regeneration of nitrogen in controlled aquatic ecosystems and the effects of copper on these processes. *Bull. Mar. Sci.* **30**: 431-443.

Houde, E. D. and S. A. Berkeley. 1982. Food and growth of juvenile herring, *Clupea harengus pallasi*, in CEPEX enclosures. Pp. 239-249. **In**: G. D. Grice and M. R. Reeve [eds.], *Marine Mesocosms. Biological and Chemical Research in Experimental Ecosystems.* New York: Springer-Verlag.

Koeller, P. and T. R. Parsons. 1977. The growth of young salmonids (*Onchorhynchus keta*): controlled ecosystem pollution experiment. *Bull. Mar. Sci.* **27**: 114-118.

Kudoh, S. 1988. Environmental controlling mechanisms on the population change of a diatom, *Asterionella formosa* Hass., in lakes. University of Tokyo: MS thesis. 97 pp. [In Japanese.]

Lee, R. F., M. Takahashi, J. R. Beers, W. H. Thomas, D. L. Seibert, P. Koeller, and D. R. Green. 1977. Controlled ecosystems: Their use in the study of the effects of petroleum hydrocarbons on plankton. Pp. 323-342. **In**: F. J. Vernberg, A. Calabrese, F. P. Thurberg, and W. B. Vernberg [eds.], *Physiological Responses of Marine Biota to Pollutants.* New York: Academic Press.

McAllister, C. D., T. R. Parsons, K. Stephens, and J. D. H. Strickland. 1961. Measurements of primary production in coastal sea water using a large-volume plastic sphere. *Limnol. Oceanogr.* **6**: 237-258.

Menzel, D. W. 1980. Applying results derived from experimental microcosms to the study of natural pelagic marine ecosystems. Pp. 742-752. **In**: J. P. Giesy, Jr. [ed.], *Microcosms in Ecological Research*. DOE Symposium Series 52, CONF-781101. Springfield, VA: National Technical Information Service.

Mullin, M. M. 1980. How can enclosing seawater liberate biological oceanographers? Pp. 399-410. **In**: G. D. Grice and M. R. Reeve [eds.], *Marine Mesocosms. Biological and Chemical Research in Experimental Ecosystems.* New York: Springer-Verlag.

Mullin, M. M. and P. M. Evans. 1974. The use of a deep tank in plankton ecology. 2. Efficiency of a planktonic food chain. *Limnol. Oceanogr.* **19**: 902-911.

Odum, E. P. 1984. The mesocosms. *Bioscience* **34**: 558-562.

Otsuki, A., M. Aizaki, T. Iwakuma, N. Takamura, T. Hanazato, T. Kawai, and M. Yasuno. 1985. Coupled transformation of inorganic stable carbon-13 and nitrogen-15 isotopes into higher trophic levels in a eutrophic shallow lake. *Limnol. Oceanogr.* **30**: 820-825.

Parsons, T. R. 1981. The use of controlled experimental ecosystems: A review. *J. Oceanogr. Soc. Japan* **37**: 294-298.

Parsons, T. R. 1988. Trophodynamic phasing in theoretical, experimental and natural pelagic ecosystems. *J. Oceanogr. Soc. Japan* **44**: 94-101.

Parsons, T. R., K. von Brockel, P. Koeller, M. Takahashi, M. R. Reeve, and O. Holm-Hansen. 1977a. The distribution of organic carbon in a marine planktonic food web following nutrient enrichment. *J. Exp. Mar. Biol. Ecol.* **26**: 235-247.

Parsons, T. R., W. H. Thomas, D. Seibert, J. R. Beers, P. Gillespie, and C. Bawden. 1977b. The effect of nutrient enrichment on the plankton community in enclosed water columns. *Int. Rev. Ges. Hydrobiol.* **62**: 565-572.

Parsons, T. R., P. J. Harrison, and R. Waters. 1978. An experimental simulation of changes in diatom and flagellate blooms. *J. Exp. Mar. Biol. Ecol.* **32**: 285-294.

Parsons, T. R., P. Thompson, Wu Yong, C. M. Lalli, Hou Shumin, and Xu Huaishu. 1986. The effect of mine tailings on the production of plankton. *Acta Oceanologica Sinica* **5**: 417-423.

Parsons, T. R. and C. M. Lalli. 1988. Comparative oceanic ecology of the plankton communities of the subarctic Atlantic and Pacific Oceans. *Oceanogr. Mar. Biol. Annu. Rev.* **26**: 317-359.

Patten, B. C. and E. P Odum. 1981. The cybernetic nature of ecosystems. *Am. Naturalist* **118**: 886-895.

Ryther, J. H. 1969. Photosynthesis and fish production in the sea. *Science* **166**: 72-76.

Sakamoto, M., A. Otsuki, M. Yasuno, T. Iwakuma, T. Hanazato, and K. Aoyama. 1988. Evaluation of a planktonic food web system in a spring diatom bloom community by the ^{13}C and ^{15}N tracer method. Pp. 114-125. **In**: Y. Saijo [ed.], *Scientific Report of the Experimental Analysis of Trophic Interactions using Mesocosms (EXATIM)*. Nagoya University: Water Research Institute. [In Japanese.]

Sonntag, N. C. and T. R. Parsons. 1979. Mixing an enclosed, 1300 m^3 water column: effects on the planktonic food web. *J. Plankton Res.* **1**: 85-102.

Steele, J. H., D. M. Farmer, and E. W. Henderson. 1977. Circulation and temperature structure in large marine enclosures. *J. Fish. Res. Bd. Canada* **34**: 1095-1104.

Steele, J. H. and J. C. Gamble. 1980. Predator control in enclosures. Pp. 227-237. **In**: G. D. Grice and M. R. Reeve [eds.], *Marine Mesocosms. Biological and Chemical Research in Experimental Ecosystems.* New York: Springer-Verlag.

Strickland, J. D. H. and L. D. B. Terhune. 1961. The study of *in-situ* marine photosynthesis using a large plastic bag. *Limnol. Oceanogr.* **6**: 93-96.

Strickland, J. D. H., O. Holm-Hansen, R. W. Eppley, and R. J. Linn. 1969. The use of a deep tank in plankton ecology. I. Studies of the growth and composition of phytoplankton crops at low nutrient levels. *Limnol. Oceanogr.* **14**: 23-24.

Takahashi, M., W. H. Thomas, D. L. R. Seibert, J. Beers, P. Koeller, and T. R. Parsons. 1975. The replication of biological events in enclosed water columns. *Arch. Hydrobiol.* **76**: 5-23.

Takahashi, M., Y. Watanabe, K. Kato, I. Yasuda, A. Otsuki, T. Hanazato, H. Hayashi, S. Araki, M. Kishino, H. Toda, S. Kudo, T. Okino, M. Sakamoto, K. Yoshida, K. Aoyama, and Y. Yamamoto. 1988. Ecosystem responses to shading perturbation using mesocosms. Pp. 27-50. **In**: Y. Saijo [ed.], *Scientific Report of the Experimental Analysis of Trophic Interactions using Mesocosms (EXATIM)*. Nagoya University: Water Research Institute. [In Japanese.]

Thomas, W. H., D. L. R. Seibert, and M. Takahashi. 1977. Controlled ecosystem pollution experiment: Effect of mercury on enclosed water columns. III. Phytoplankton population dynamics and production. *Mar. Sci. Comm.* **3**: 331-354.

5. PELAGIC MESOCOSMS: II. PROCESS STUDIES

U. Brockmann

Abstract

Examples of ecological research using pelagic mesocosms are reviewed in order to show the applicability of experimental ecosystems in studying specific ecological processes. There are still shortcomings in the application of this specific tool in biological oceanography. Many process studies have been performed in mesocosms such as transfer and transformation of chemical compounds, and rate measurements of biological and chemical processes such as: (1) primary production; (2) release of organic compounds by phytoplankton; (3) decomposition measured as respiration or release of ammonium and urea; (4) grazing at different food size spectra; and (5) sedimentation rates, which could be related to the total development of the enclosed ecosystem. Within deep enclosures, vertical processes were studied, such as migration and the boundary effects of densiclines. By frequent sampling, diurnal and subdiurnal behavior were investigated. The succession of phytoplankton species and zooplankton cohorts, including the interaction between trophic levels, were studied in experiments lasting from several weeks to months. Fluxes of material can be followed by frequent measurements of the different compartments as well as by the use of isotope-spiked preconditioned organisms or nutrients. A more effective interdisciplinary utilization of mesocosm research is recommended. Evaluation of representativity can be achieved by combination with parallel field measurements. The supplementation with simplified systems (monocultures) and laboratory experiments can contribute significantly to ecosystem analysis.

INTRODUCTION

The statement noted by Gamble and Davies (1982) in their review is still valid: that one of the main advantages of mesocosm experiments is to sample the same multitrophic ecosystem in the same body of water over more than the generation time of at least the third trophic level.

In comparison to field studies or laboratory experiments, mesocosm research is especially qualified for the investigation of:

(1) Transfer and transformation of chemical substances, including turnover rates and fluxes of material based on budgets. This includes not only chemical reactions and physico-chemical sorption processes but also the complex field of biological processes such as nutrient uptake (especially primary production), and conversion of nutrients into organic compounds or other nutrients by nitrification, release of compounds, and decomposition.

(2) Effects of chemical compounds, e.g. by nutrient limitation, eutrophication, and pollution, including the entire range from inhibition to synergistic effects and formation of toxins in the marine environment.

(3) Succession of plankton groups, especially of phytoplankton (metabolic phases, species, and size spectra) and copepod cohorts.

(4) Interaction of trophic levels (grazing, enhancement, and inhibition).

(5) Vertical processes like migration, diffusion, and sedimentation; vertical differences in turnover and behavior of organisms.

(6) Diurnal dynamics and biological rhythms.

(7) Validation of numerical models.

(8) Test of laboratory results under "field conditions".

Mesocosm experiments are suitable for validation of rate measurements which are performed by small-scale (bottle) incubation experiments and which often do not match field observations, e.g. measurements of primary production including release of dissolved organic substances. Mesocosms also facilitate the comparison and intercalibration of different methods (e.g. ^{14}C, O_2, pH, carbon increase, and estimation of growth rates by cell counts).

The wide scope of mesocosm research requires a careful set-up of hardware and interdisciplinary field work in order to use effectively the excellent possibilities offered by enclosure systems individually designed for specific, ecological research programs.

Banse (1982), critically looking at mesocosm research with historical hindsight, pointed out that this type of work requires a large personnel budget in order to utilize the facilities effectively and to answer the many open questions in aquatic research. He criticized that:

(a) facilities often were not used intensively enough (in some cases, only preliminary studies were performed);

(b) zooplankton dynamics scarcely have been investigated, especially in regard to the fact that animal populations in marine enclosures were not manipulated in order to study predation effects on zooplankton composition;

(c) detailed budgets of biogenous elements were rarely performed;

(d) numerical models, including mass balances, were rarely applied and were too seldom used for conception of experiments (e.g. minimum number of parameters, sampling frequency); and

(e) interactions between plankton succession and benthic processes were not studied.

Banse assumed that processes in semi-natural plankton populations may be still too complex for testing biological models, and he suggested using synthetic systems of well-known species. He pointed out: "If we cannot model bags or beakers, how will we ever understand the ocean?"

Some reasons for the shortcomings in mesocosm research are listed below.

(a) Budgets have been too small and the duration of funding too short.

(b) Zooplankton sampling requires larger quantities of water which cannot be provided by smaller enclosures without manipulating the enclosed system significantly.

(c) For budgets, at least the estimation of biogenous elements (C, N, P, Si) in the different phases (dissolved inorganic, organic and particulate) is needed as well as rate measurements of external sources and sinks (CO_2-uptake, respiration, N_2-fixation, denitrification, sedimentation, remobilization and resuspension). The facilities for even these basic measurements are very seldom available. Some methods (e.g. ^{14}C, estimation of dissolved organic compounds) are still not standardized, thus lacking in necessary precision.

(d) Interdisciplinary cooperation between scientists of different disciplines, such as biology and mathematics, can only be successfully achieved within a longer time of cooperation (i.e. long-time funding is necessary).

(e) Mesocosm experiments are mainly designed for benthic or pelagic research. A combined system will not only increase the overall complexity but also requires an increase in the number of cooperating specialists necessary. Cooperation within such a large group of benthologists and planktologists, including the different specialists for chemistry, physical parameters, phytoplankton (including small flagellates), bacteria, zooplankton, meiobenthos, macrobenthos, sediments, and particle spectra, can only be successful if an accepted senior scientist with an interdisciplinary understanding is available, or within a partnership cooperation in which the team members are trained prior to the experiment. Generally, in interdisciplinary cooperation, it is less worthwhile to estimate single components by very sophisticated methods, which will only benefit the fame of a researcher specialist, than to analyze high numbers of group or sum parameters of general importance that are needed for understanding of the basic processes of the enclosed ecosystem.

In addition to Banse's (1982) remarks, some other critical points about mesocosm research can be mentioned. These include the facts that:

(a) analyses of bacteria and micro-flagellates were often not adequate;

(b) comparisons of enclosure results with parallel field measurements (evaluation of representativity) were rarely performed;

(c) combination of natural enclosed ecosystems with simplified systems (for instance, rate measurements of isolated dominant or relevant species in laboratory-scale experiments) was not used intensively enough to explain complex processes; and

(d) measurements of important chemical parameters, which should have been known in order to understand such things as succession processes (trace metals, vitamins), were omitted.

The above criticisms may also apply to many field investigations but, in contrast to limitations of staff and equipment on board a research vessel, there are no limitations for any convenient analysis in mesocosm research, and advection will not interfere with ecological changes.

In spite of the listed shortcomings, meaningful results have been obtained from pelagic mesocosms that could not have been revealed so effectively either by extended field work or by laboratory experiments with their limited representativity. The reproducibility of enclosure experiments has been proved for several systems (Takahashi et al., 1975; Brockmann et al., 1977a; Kuiper, 1977, 1981; Smith et al., 1982; Takahashi, this volume), but especially large mesocosms cannot be controlled completely nor precisely repeated like laboratory experiments (Reeve et al., 1982).

In this paper, some selected results of process studies will be discussed that show the potential of pelagic mesocosm research.

TRANSFER AND TRANSFORMATION OF CHEMICAL COMPOUNDS

Transfer and transformation of individual substances can be followed best within enclosures. Therefore, mesocosm systems are frequently used for pollution research; realistic transfer rates of pollutants between the different phases (dissolved and particulate), as well as between different trophic levels, can be obtained. Furthermore, the transformation of pollutants, e.g. the formation of methyl mercury, can be quantified under near-natural conditions. Examples of pollution research using mesocosms are given by Pilson in this volume.

Effects of different levels of nutrient addition have been investigated in many mesocosm experiments. Oviatt et al. (1986a), using the MERL-system for a two-year period, studied the effects of daily nutrient addition of different concentrations simulating discharge effects along an estuary. The nutrient additions resulted mainly in a higher respiratory demand of the organisms in the water column than in the benthos. In some cases, temporary anoxic conditions in the aqueous phase were observed. The transfer and transformation of biogenic elements is described in more detail in the following sections (cf. fluxes of material).

RATE MEASUREMENTS OF BIOLOGICAL AND CHEMICAL PROCESSES

Primary Production

Already in the first plastic sphere experiments, several methods of primary production measurements were compared (McAllister et al., 1961; Gamble and Davies, 1982). Davies and Williams (1984) found that ^{14}C measurements were not equivalent to oxygen-based productivity estimations. Oviatt et al. (1986b) found that, within long-term analysis, the results revealed by the ^{14}C method were less comparable than those by O_2 and CO_2 measurements.

The use of an incubation technique (e.g. ^{14}C) in small-scale experiments run parallel to mesocosms is probably the most widely used combination of micro- and mesoscale experiments. Enclosures with monocultures of phytoplankton species or with low grazing pressure are excellent tools for the investigation of the still unresolved problem associated with the release of dissolved organics during primary production. It should be noted that released labile organic compounds with high nutritional value will be decomposed very rapidly by bacteria. Continuous measurements of dissolved carbohydrates at short time intervals showed that release rates were similar to decomposition rates (Brockmann et al., 1977a).

Release

Pelagic mesocosm experiments cannot be performed with axenic cultures, therefore bacterial uptake of dissolved organic substances will always occur, and frequent measurements are needed to estimate, at least, net release ratios. In enclosure experiments with monocultures of *Thalassiosira rotula* and natural bacterial populations, a release of 14 to 28% of the particulate primary production was measured for dissolved carbohydrate-carbon only (Brockmann, 1990). This release increased during the stationary phase to more than 50% (600 μg carbohydrate-carbon l^{-1} d^{-1}) (Fig. 1). In these experiments, a significant correlation between release activity and phased cell division was found. The release mainly occurred during the assimilation phase and also during periods of minimum cell divisions (generation time, 11.7 h).

During long-term experiments, wall effects, such as adsorption and uptake of dissolved organic substances by wall-fixed bacteria, have to be considered and included in decomposition estimates.

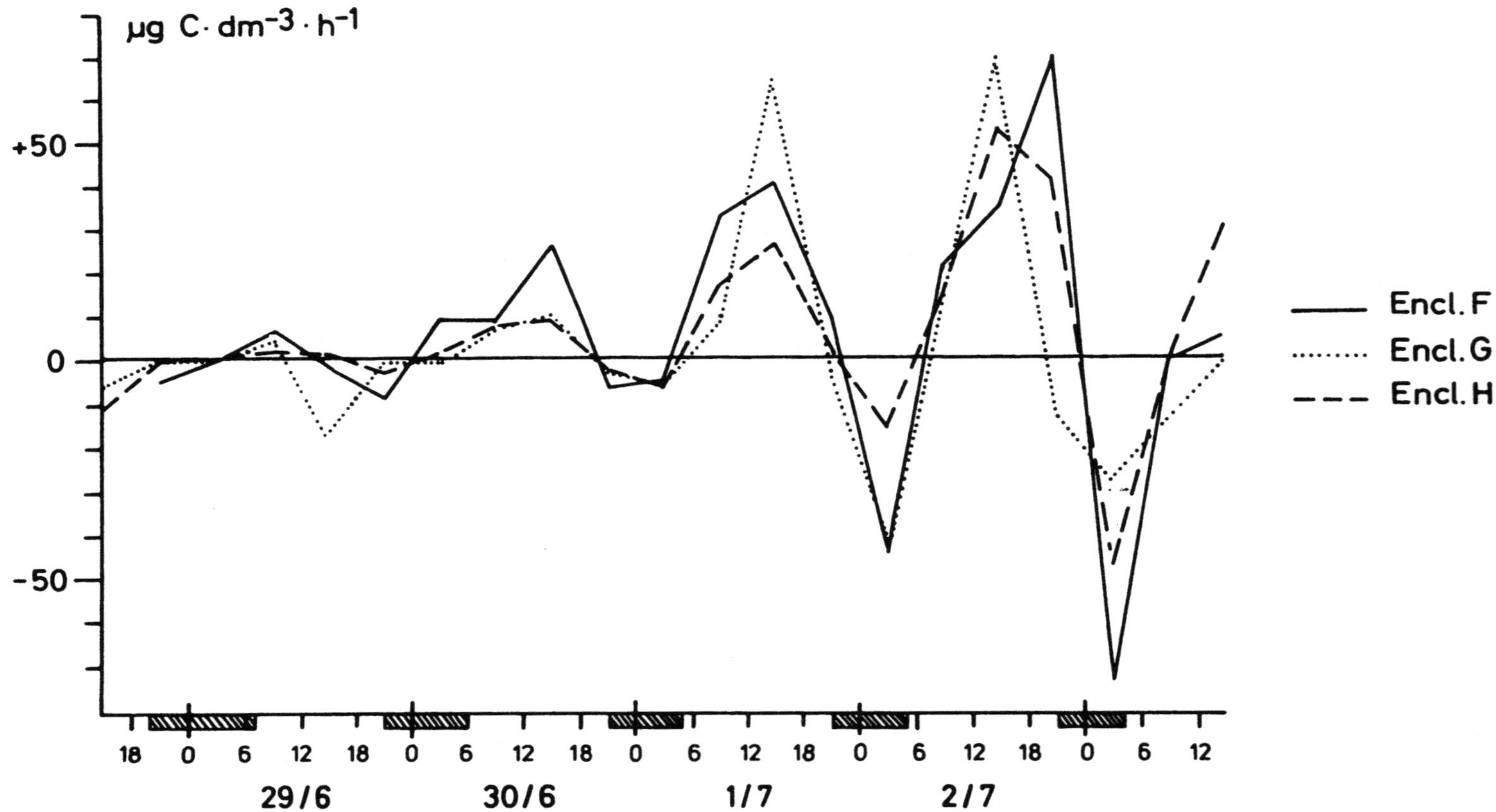

Figure 1. Diurnal concentration changes of total carbohydrates (TCH) in three parallel enclosures (F, G, H) with monocultures of *Thalassiosira rotula*. Rates of increase and decrease were calculated as the deviation of concentrations estimated every 6 hours from running means. Modified from Eberlein and Brockmann (1986).

Decomposition

Decomposition, mainly estimated by O_2-consumption, can be calculated in mesocosm experiments from production rates of heterotrophic organisms in combination with data from incubation experiments or from integrated flux studies. Ammonium and urea are the first remineralization products released after hydrolysis of proteins. Koike et al. (1982) found that, in enclosures with coastal water communities, ammonium production reached 60% of ammonium consumption.

Respiration has been calculated by Williams (1982) using size-dependent equations based on the biomass of organisms and empirical constants for the different size groups. These calculations were in good agreement with respiration measurements (O_2-consumption) of different size-classes, revealing that 50% of respiration was associated with organisms smaller than 1 μm.

Grazing

A direct comparison of grazing measurements in grazing chambers using labelled food with studies of food webs in enclosures have probably not been performed up to now. On the other hand, grazing has been estimated in combination with mesocosm experiments by using the clearance rate of isolated zooplankton of known composition in bottles (Harris et al., 1982). Individual rates of isolated single zooplankton species in small containers were combined with the development of numbers and size classes of animals in enclosures in order to calculate ingestion rates. In these experiments, the effects of species and food-size dependent grazing also was studied with isolated single zooplankton species in small containers.

Ingestion rates of adult females of *Pseudocalanus* and *Calanus* feeding on phytoplankton showed large differences for *Calanus* spp. between two bags, probably due to different food-size spectra. The grazing rate was higher in an enclosure in which the particle spectra showed a maximum above 50 μm, whereas particles were mainly below 30 μm in the other enclosure (Fig. 2). Small copepods like *Pseudocalanus*, which preferentially graze on small particles, showed no significant differences in ingestion rates.

Sedimentation

Increased sedimentation, especially of diatoms, was often observed within enclosures (with different turbulence in comparison to the environment). Von Bröckel (1982) measured sedimentation rates of more than 140 mg C m^{-2} d^{-1} (trap material) after launching the bags (Fig. 3). This was assumed to be an effect of reduced turbulence in the enclosure, because these bags were designed mainly for long-term investigations of higher trophic levels and were therefore constructed of multiple layers of plastic, reducing

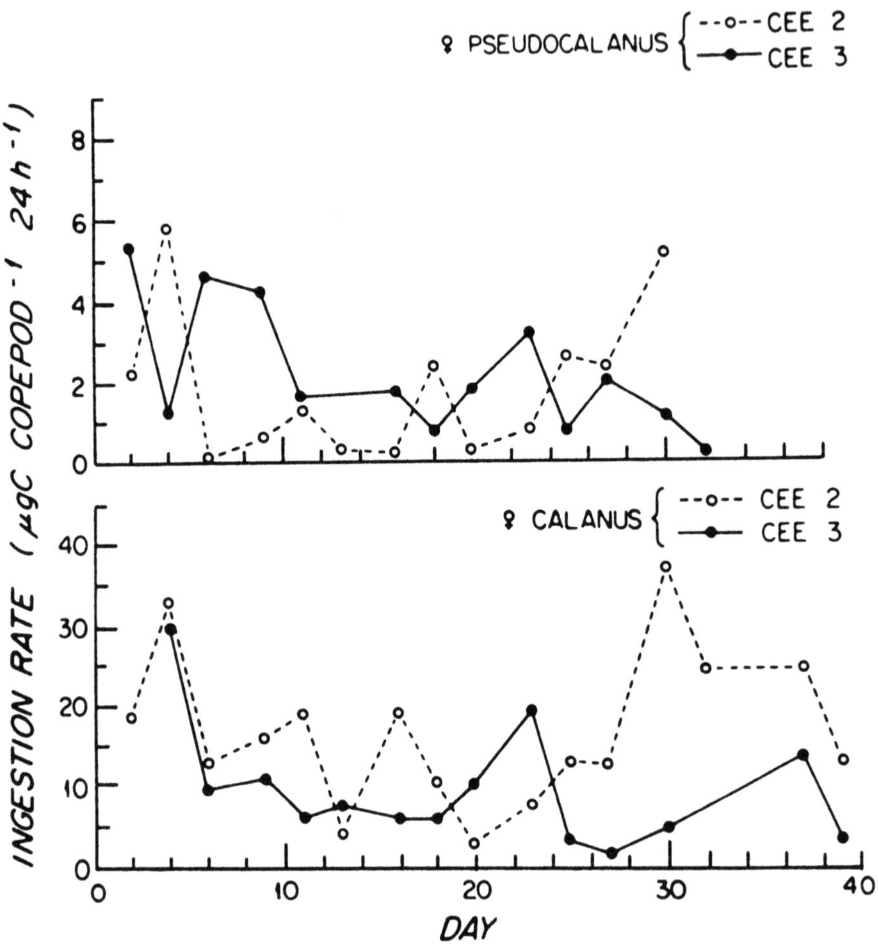

Figure 2. Ingestion rates of adult female herbivorous copepods feeding on phytoplankton obtained from 4-8 m depths in two mesocosms (CEE 2 and 3). CEE 2 was intermittently mixed by bubbling; incident light was reduced by a screen, causing a phytoplankton dominance with a particle size maximum above 50 μm following day 9. From Harris et al. (1982).

energy transfer through the walls. In order to achieve sufficient turbulence, the design of mesocosm experiments (in terms of flexibility of walls, type of stirring or air bubbling) is of utmost importance and will significantly influence sedimentation behavior of phytoplankton. On the other hand, increased sedimentation following phytoplankton blooms was observed in the open sea (Cadée, 1986). Sinking rates in the enclosures (0.6 to 5 m d^{-2}) were in the same range as those of field observations. The reduced sedimentation after the second week of the enclosure experiment was probably due to phytoplankton succession (i.e. the appearance of motile flagellates). On the

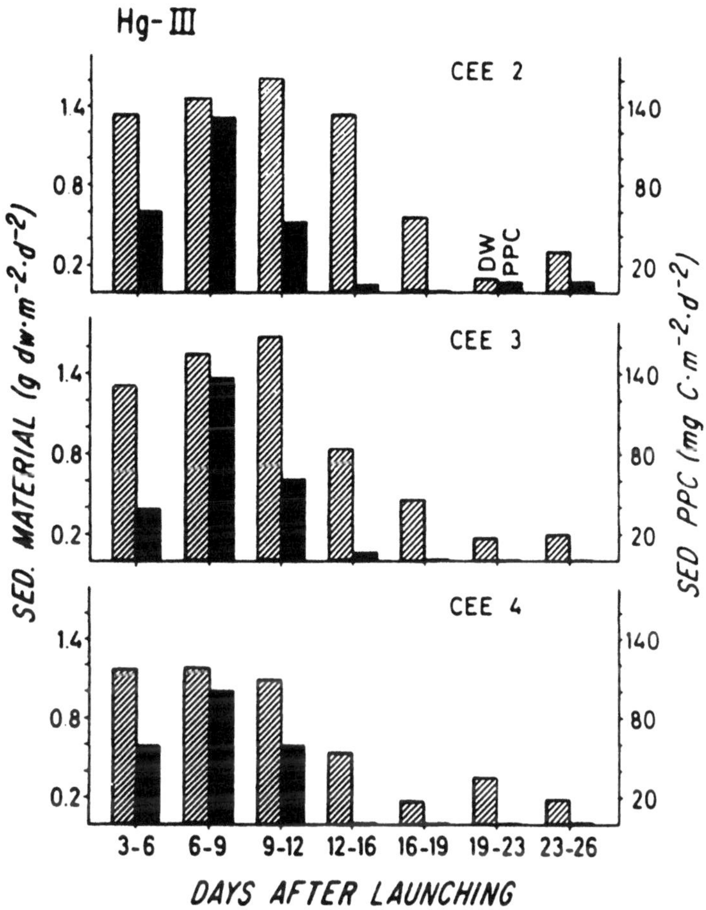

Figure 3. Sedimentation of particulate matter [as dry weight (dw) g m^{-2} d^{-1}] and phytoplankton carbon (PPC; g C m^{-2} d^{-1}) during the Hg III experiment from three enclosures (CEE 2, 3, and 4). CEE 4, control; CEE 3, 1 μg Hg l^{-1} added; CEE 2, 5 μg Hg l^{-1} added. From von Bröckel (1982).

contrary, Bienfang (1982) measured transit times of labelled cells and found low sinking rates of small, colony-building diatoms (*Chaetoceros constrictus*) following bag launching.

One general aspect has to be considered in interpreting ecological events: synoptic events may be correlated significantly without any coupled reactions. A lag phase of physiological response could have increased sedimentation rates in the above-mentioned experiment (von Bröckel, 1982); cell buoyancy may have been reduced as a consequence of nutrient limitation prior to launching of the bags. It is recommended that the state of the ecosystem be carefully characterized before starting mesocosm experiments with natural communities.

The effect of nutrient concentrations on sedimentation rates was confirmed by Bienfang (Fig. 4). Sinking rates were reduced after nutrient addition, and periods of bright light also seemed to reduce the sinking behavior of phytoplankton when *Ceratium fusus* was dominant. The lack of significant correlation between these events and sinking rates was regarded as a retarded response of cell metabolism.

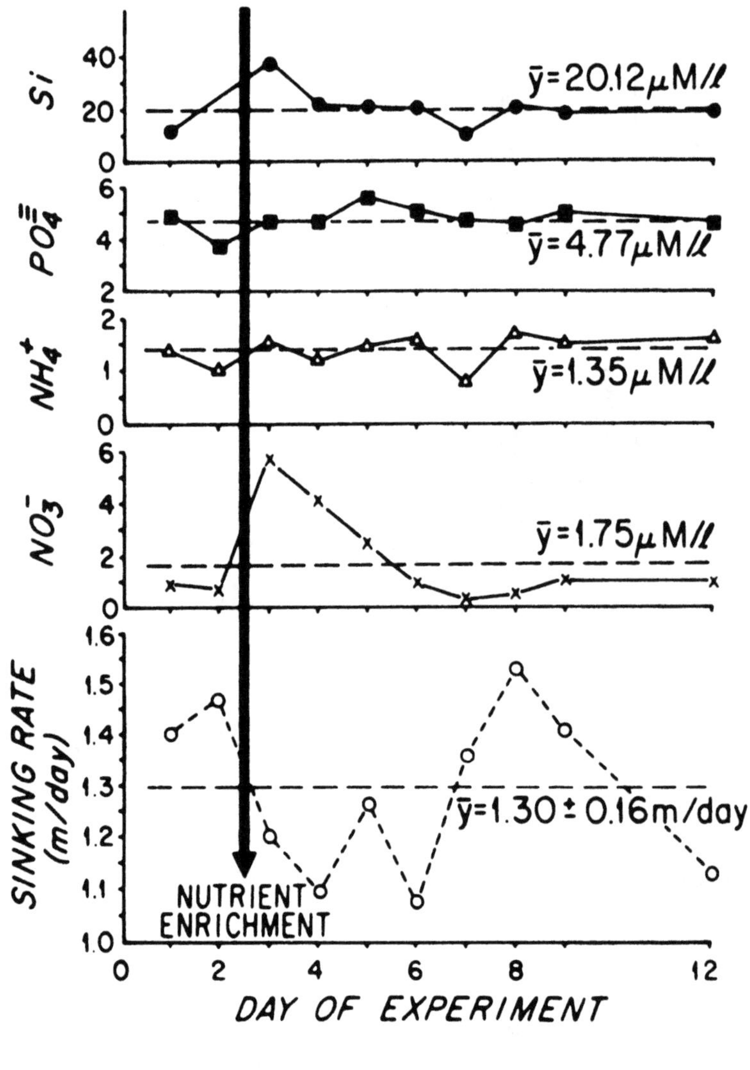

Figure 4. Changes in sinking rate of the 8-53 μm phytoplankton fraction and prevailing environmental parameters in a CEE during a 12-day experiment. Integrated samples were taken over the 0-8 m depth interval. Sinking rates represent the means of triplicates; the average coefficient of variation (SD/x) was 9%. Nutrient concentrations in μM l^{-1}. From Bienfang (1982).

VERTICAL PROCESSES

Vertical processes, like surface heating, cooling, and wind-induced turbulence, control the formation and dynamics of the mixed layer and thus the growth conditions of phytoplankton and closely interacting foodweb organisms. Turbulence is one of the essential factors keeping the nonmotile plankton floating, and it promotes the exchange of vicinal water (exhausted of nutrients and vitamins) and the dispersion of dividing cells.

Mesocosm experiments were very often performed in enclosures with flexible walls, or else turbulence was artificially introduced. Gust (1977), using flexible and small (<1 m^3) plastic bags, found comparable turbulence inside and outside (measured by hot-wire anemometry). Steele et al. (1977) investigated diffusion in large (>10 m^3) plastic enclosures (by dye experiments and thermistor chains); they found that some mixing forces were inhibited (wind stress) and others arose from the flexible wall itself (wave action). Thermal forcing in the surrounding water will be transferred to the water enclosed by plastic bags with only a short delay, dependent on the qualities of the material used. Thermal stratification can be conserved or changed in enclosures as well as in the environment. Only haline stratification guarantees stability in enclosure systems over longer time periods (Brockmann and Hentzschel, 1983).

The ability to migrate vertically gives flagellates an ecological advantage when nutrients are limited within the mixed layer. This is one of the reasons for the succession of diatoms by flagellates; the latter can form monospecies blooms, especially in coastal areas, which cause oxygen depletion after decomposition and sedimentation following nutrient depletion. Some species can cause shellfish poisoning (PSP, DSP) or fish-kills. Enclosure experiments facilitate the investigation of vertically migrating "red tide" species.

Diurnal vertical migration was observed in the species *Heterosigma akashiwo*, which causes red tides in Osaka Bay. The migration showed both negative and positive phototaxis as the algae moved to medium light levels; vertical migration was in phase with cell division (Kohata and Watanabe, 1986). The toxic *Gyrodinium aureolum* showed no diurnal vertical migration but was most abundant at the nutricline following nutrient depletion at the surface (Brockmann et al., 1985; Dahl and Brockmann, 1985, 1989).

Vertical migration of *Daphnia* in relation to chlorophyll *a* distribution was intensively studied in large lake mesocosms (George, 1983). Incident light, turbidity, and population density (food limitation) were the main factors affecting vertical distribution of *Daphnia*.

DIURNAL AND SUBDIURNAL DYNAMICS

Diurnal response of phytoplankton to the light regime has been observed in the field, and many investigations dealing with this matter have been performed in the laboratory. The interaction of different processes like assimilation rate, cell growth, qualitative and quantitative changes of cell contents, and release of cell products can be observed best under nearly natural conditions in mesocosm experiments.

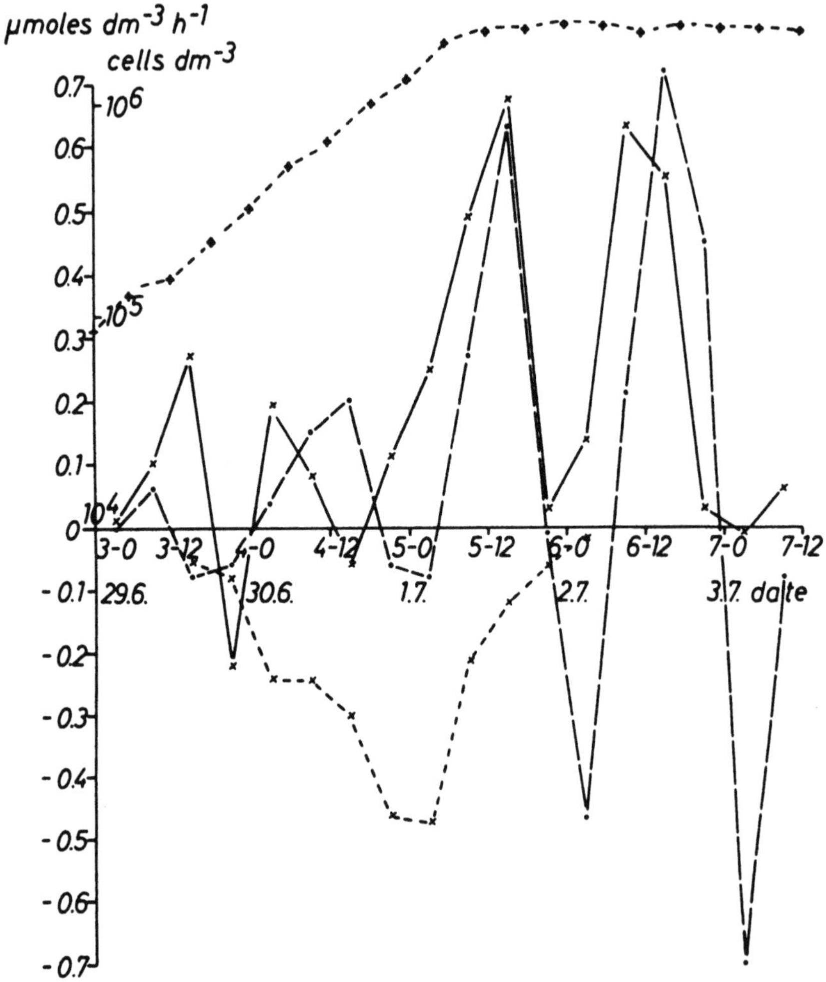

Figure 5. *Thalassiosira rotula*: development of cell numbers (+--+), rates of increase and decrease of silicate (x--x), total particulate carbohydrates (x—x), and total dissolved carbohydrates (.—.). (Mean values of three enclosures F, G, and H are given.) Rates were calculated as deviations from running means. From Eberlein and Brockmann (1986).

Phased cell division, which has been observed in open water, was achieved with *Thalassiosira rotula* monocultures in mesocosms. In these experiments, diurnal changes in particulate carbohydrate as well as in dissolved carbohydrates were found (Eberlein and Brockmann, 1986) (Fig. 5). Maximum release occurred during daytime at maximum assimilation activity. Maximum carbohydrate concentrations were reached within the cells just before maximum release of carbohydrates occurred, indicating that a release immediately followed the formation of biomass. Both concentrations were similar. During the stationary phase, release rates reached more than 0.5 μM dm^{-3} h^{-1}. The synchronized growth rate with two divisions per day (Brockmann et al., 1977a) caused phased release signals, especially for free amino acids (Hammer and Brockmann, 1983) (Fig. 6). These observations show the close relationship between release and decomposition. Chelation of heavy metals or symbiotic relations to vitamin-producing bacteria are discussed as reasons for the high release rates.

SUCCESSION

Succession of Phytoplankton Metabolic Phases

Different metabolic phases of phytoplankton development are defined from batch cultures: lag phase, exponential growth phase, stationary and decomposition phases. These phases can also be observed in the field during the spring bloom and after upwelling events. The changes in cell metabolism are difficult to detect in the field due to many interfering reactions. Based on enclosure experiments, the increase of phytoplankton C/N ratios at the stationary growth phase was described by Antia et al. (1963). The changes in cellular compounds as well as in release products could be observed in enclosures with diatom monocultures (Brockmann et al., 1977a). Primary production did not cease following depletion of nitrogen (Fig. 7), and this resulted in a doubling of carbohydrate cell content. Release of carbohydrates increased concurrently in relation to primary production, and the composition of carbohydrates changed to mainly storage products (polymers of glucose).

The changes in composition of released free amino acids caused a shift of C/N ratios to higher values during the stationary phase. This was observed in the field (FLEX) as well as in enclosures (Hammer et al., 1981; Hammer and Kattner, 1986).

Succession of Phytoplankton Species (including Size Spectra)

A summary of factors influencing phytoplankton selection has been given by Harrison and Turpin (1982) (Table 1).

Temperature optima for different groups and species range from 0° to 40^oC. At low temperatures, temperature-dependent succession can

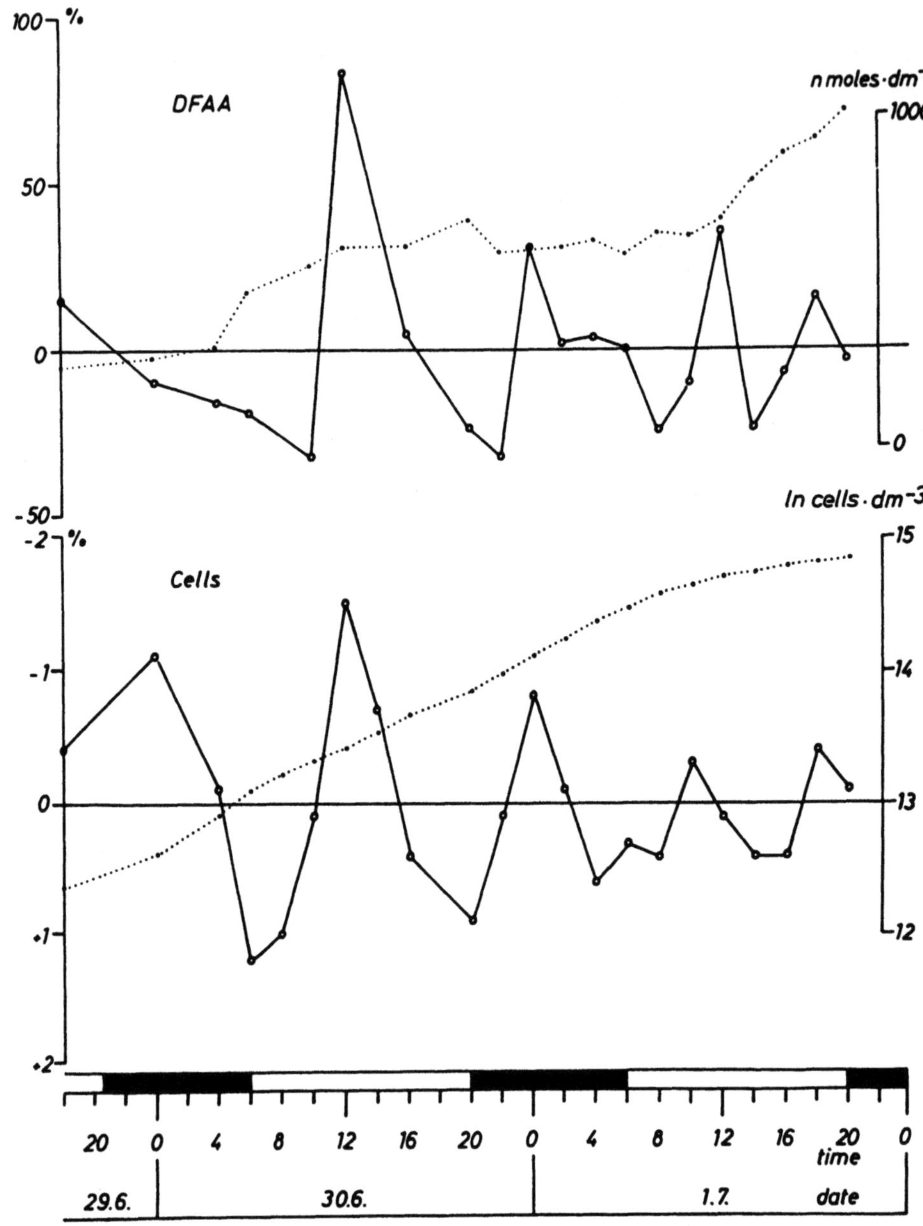

Figure 6. Periodic short-term fluctuations of dissolved free amino acids (DFAA) and logarithmic cell counts of *Thalassiosira rotula*, presented inversely as deviation from the trend in percent (straight line) and long-term trend values (dotted line), in nmoles dm^{-3} and logarithmic cell numbers dm^{-3}, respectively (enclosure H, cf. Fig. 1). From Brockmann et al. (1977a) and Hammer and Brockmann (1983).

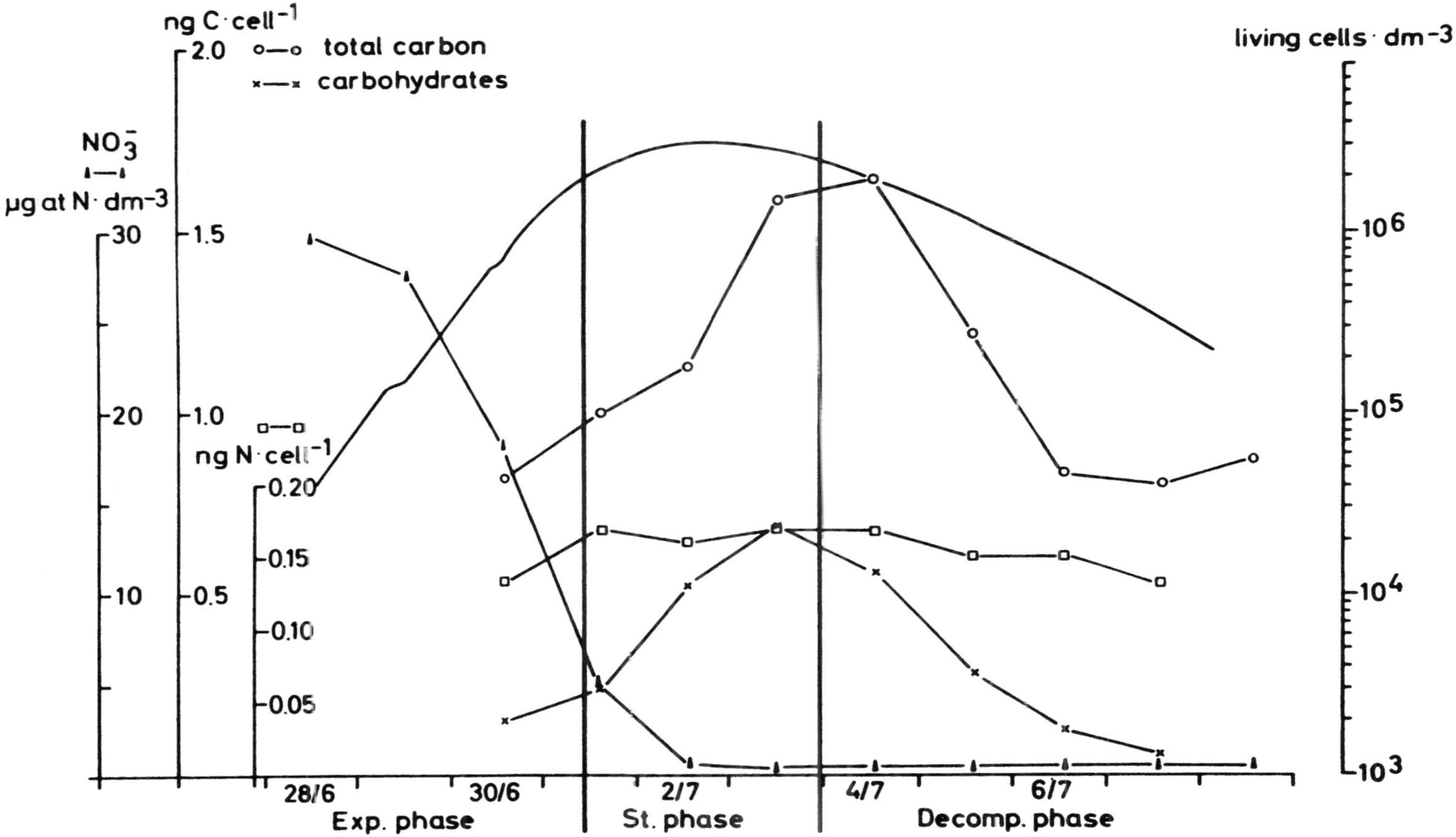

Figure 7. Particulate components of monoculture experiments (means of three enclosures) with *Thalassiosira rotula*: increase of particulate carbon during the stationary growth phase. From Brockmann (1990).

Table 1.

A summary of factors arranged according to their capability of selecting among phytoplankton groups (coarse tuning), within groups (fine tuning), or different cell sizes. From Harrison and Turpin (1982).

I. Coarse tuning factors
   1. Temperature
   2. Specific nutrient flux
   3. Nutrient ratio (N:Si or P:Si)
   4. Irradiance

II. Fine tuning factors
   1. Frequency of addition of the limiting nutrient
   2. Form of the limiting nutrient
   3. Allelopathy
   4. Parasitism

III. Cell size selection factors
   1. Sinking
   2. Temperature
   3. Frequency of addition of the limiting nutrient
   4. Size-selective grazing

sometimes be observed or even temperature-dependent primary production (Jahnke et al., 1983).

The specific nutrient flux (ratio of nutrient uptake and concentration of nutrient elements per hour) strongly influences the species composition in mesocosm experiments. High specific nutrient fluxes enhance small, fast-growing centric diatoms, whereas low specific nutrient fluxes favor mixtures of small flagellates, pennate diatoms and larger, slower-growing diatoms (see also Davis, 1982 and Takahashi et al., 1982). Nutrient ratios also affect species dominance. This has mainly been documented in freshwater studies (Schindler, 1977).

Cell size and cell biomass are at a minimum at temperature optimum in order to compete successfully at nutrient limitation (Goldman and Mann, 1980). A decrease of the pervalva axis was observed in enclosure experiments with monocultures of *Thalassiosira rotula* and *Skeletonema costatum* (Brockmann et al., 1977a) (Fig. 8). Phased cell divisions co-occur with phased alterations of chain lengths, resulting in changes of particle size spectra (a combination of cell size and chain length) during exponential growth. These modifications will affect plankton succession with respect to differences in sinking behavior or size-selective grazing.

The silicate limitation of diatoms after the spring bloom is well known in temperate areas and has been observed in several enclosure experiments. The competition of fast-growing diatoms, which are preferentially grazed by herbivores in contrast to toxic dinoflagellates, will probably be an important topic of future mesocosm research. Due to the increase in reports of toxic phytoplankton blooms, selection criteria for toxic species will become still more important. Brockmann et al. (1985) observed an increase of diatom species in the presence of a *Gyrodinium aureolum* bloom, whereas other species like *Chrysochromulina polylepis* (E. Dahl, pers. commun.) may inhibit the growth of diatoms.

Temperature and nutrient limitation are among the main factors causing succession in the field, in addition to size-selective grazing and allelopathy (inhibition or stimulation by released substances). Vitamins or trace elements can influence species succession as well and sometimes cause the formation of group successions (Brockmann et al., 1977b).

When no dominating factor can be identified, it is difficult to understand an observed succession of events even in enclosed systems. Parsons et al. (1978) combined light and nutrient effects and developed a general hypothesis for growth optima of phytoplankton groups. Progress in these matters can be achieved by combining physiological studies with mesocosm research. Numerical modelling will be a valuable tool in testing succession hypotheses.

Succession of Zooplankton Cohorts

At low, natural, zooplankton concentrations, the development of copepod cohorts (Beers et al., 1977; Harris et al., 1982) as well as the succession of zooplankton size-classes can be studied in enclosures to give a realistic data base for production calculations in the field. The development of *Paracalanus* from the first naupliar stages to adult animals required 15 days (Harris et al., 1982) (Fig. 9). The mortality due to predation can be quantified from the decrease in numbers of succeeding cohorts.

The succession of herbivorous zooplankton is mainly determined by predation. Harris et al. (1982) followed the development of copepod cohorts and also of size-classes of different carnivores and calculated the energy transfer between the different trophic stages. The decrease of herbivores was generally a consequence of an increase in large predator stages; as a further consequence, phytoplankton biomass will increase.

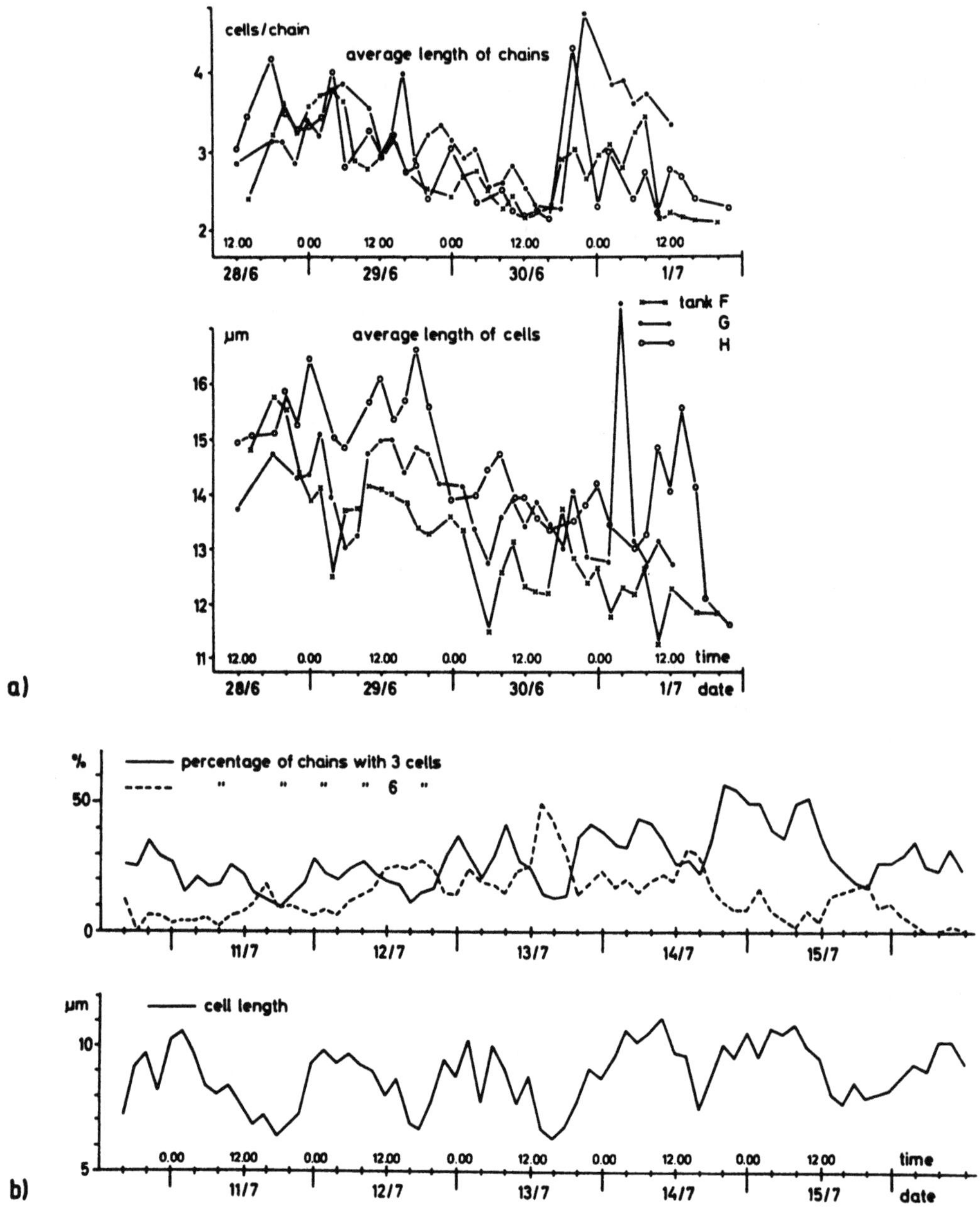

Figure 8. Succession of cell size and chain length of diatoms. (a) Average chain length (cell number) and average length of pervalvar axis of *Thalassiosira rotula*, tanks F, G, and H. (b) Length of *Skeletonema costatum* cells (pervalvar axis) and percentage of cell chains with 3 and 6 cells. From Brockmann et al. (1977a).

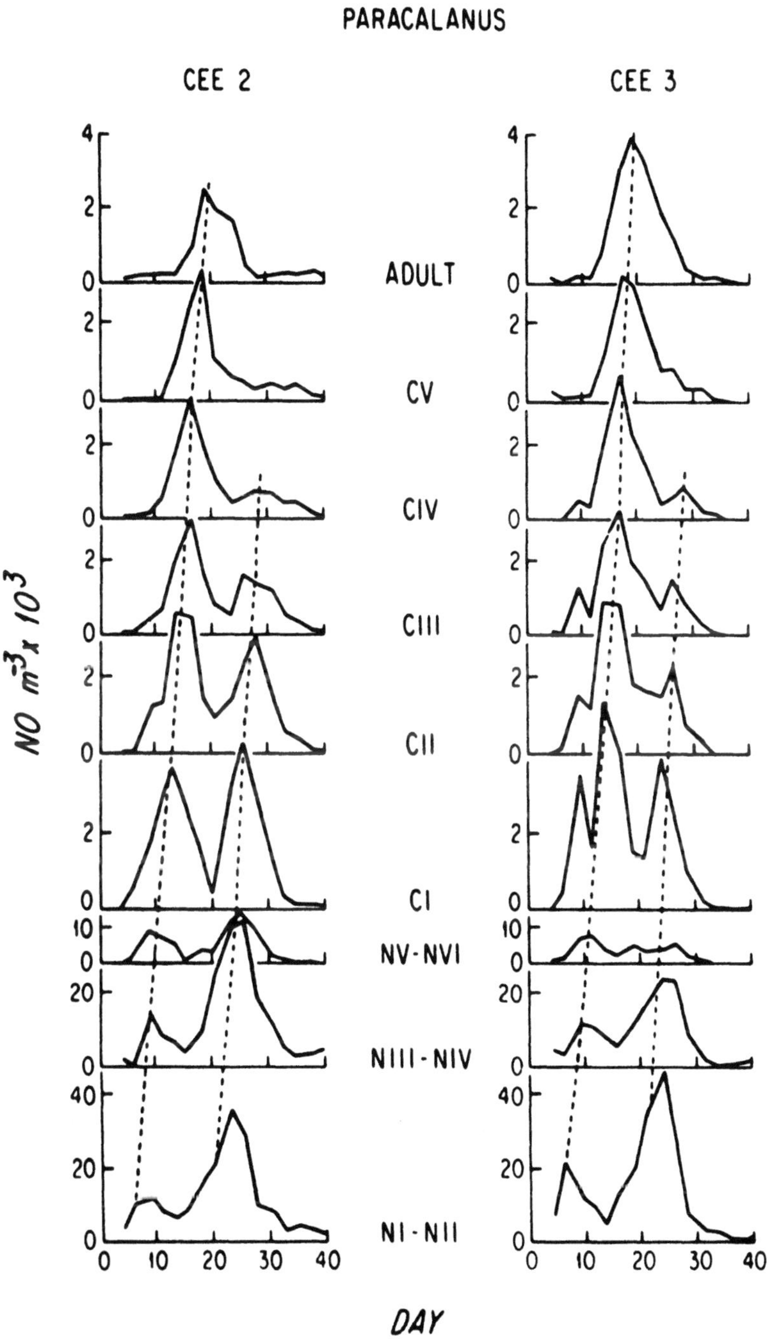

Figure 9. *Paracalanus* abundance in CEE 2 and 3 (cf. Fig. 2) by developmental stages. N represents naupliar stages with NI-NII, etc., combined; C represents copepodid stages. Note scale change between nauplii and copepodids. From Harris et al. (1982).

INTERACTIONS BETWEEN TROPHIC LEVELS

Steele and Gamble (1982) summarized the interactions between predators, herbivores, and phytoplankton in enclosure experiments. High numbers of predators produced low herbivore populations and large phytoplankton cells. Conversely, low numbers of predators led to more herbivores and small phytoplankton cells. The large or small phytoplankton cells may be diatoms or dinoflagellates. During phytoplankton succession, a shift from smaller to larger species due to differential grazing was seen in two enclosure experiments (Fig. 10) (Steele and Gamble, 1982). First, the diatom *Leptocylindrus minimus* (about 6 μm diameter) dominated; later, in the presence of higher predator concentrations, *Gyrodinium aureolum*, a toxic dinoflagellate of about 20 μm diameter, became abundant. These indirect effects of predator concentrations on phytoplankton size spectra were generally confirmed by field observations (Gamble, 1978). Similar results from enclosure research were obtained by Hessen and Nilssen (1986) in a eutrophic lake. Predation was also found to be the dominant factor controlling zooplankton community structure in a mercury-polluted clay lake (Salki et al., 1985). Many mesocosm experiments with pollutants reveal results which show mechanisms of interactions between trophic levels. One example was the higher abundance of larger phytoplankton species in polluted bags as a consequence of reduced herbivorous zooplankton (Kuiper and Gamble, 1988). Another example was an apparent increase, in relation to the control, in omnivorous copepods at low cadmium concentrations; but the real effect was the decrease of *Pleurobrachia pileus* density from the added pollutant, resulting in reduced predation (Kuiper, 1981). Food chain efficiencies are discussed by Takahashi elsewhere in this volume.

FLUXES OF MATERIAL

The flux of nitrogen in enclosures was roughly estimated by use of ^{15}N-NO_3 and ^{15}N-NH_4 which had been added to the total enclosed system (Harrison and Davies, 1977). By weekly measurements after nutrient enrichment, they observed high settling rates of particle-bound nitrogen, which they referred to as "enclosure effects". No information was given concerning particle formation during grazing, or on phytoplankton composition, and on other nutrients, e.g. silicate. Thus the observed development may also reflect eutrophication effects. More frequent measurements would have given more insight into the fluxes.

Cycling of nitrogen was also investigated with ^{15}N-NH_4 spiking by Koike et al. (1982). By more frequent measurements, they achieved a sufficient flux control which was supported by rate measurements. The steady decrease of labelled ammonia in the mixed layer, parallel to an increase of particulate

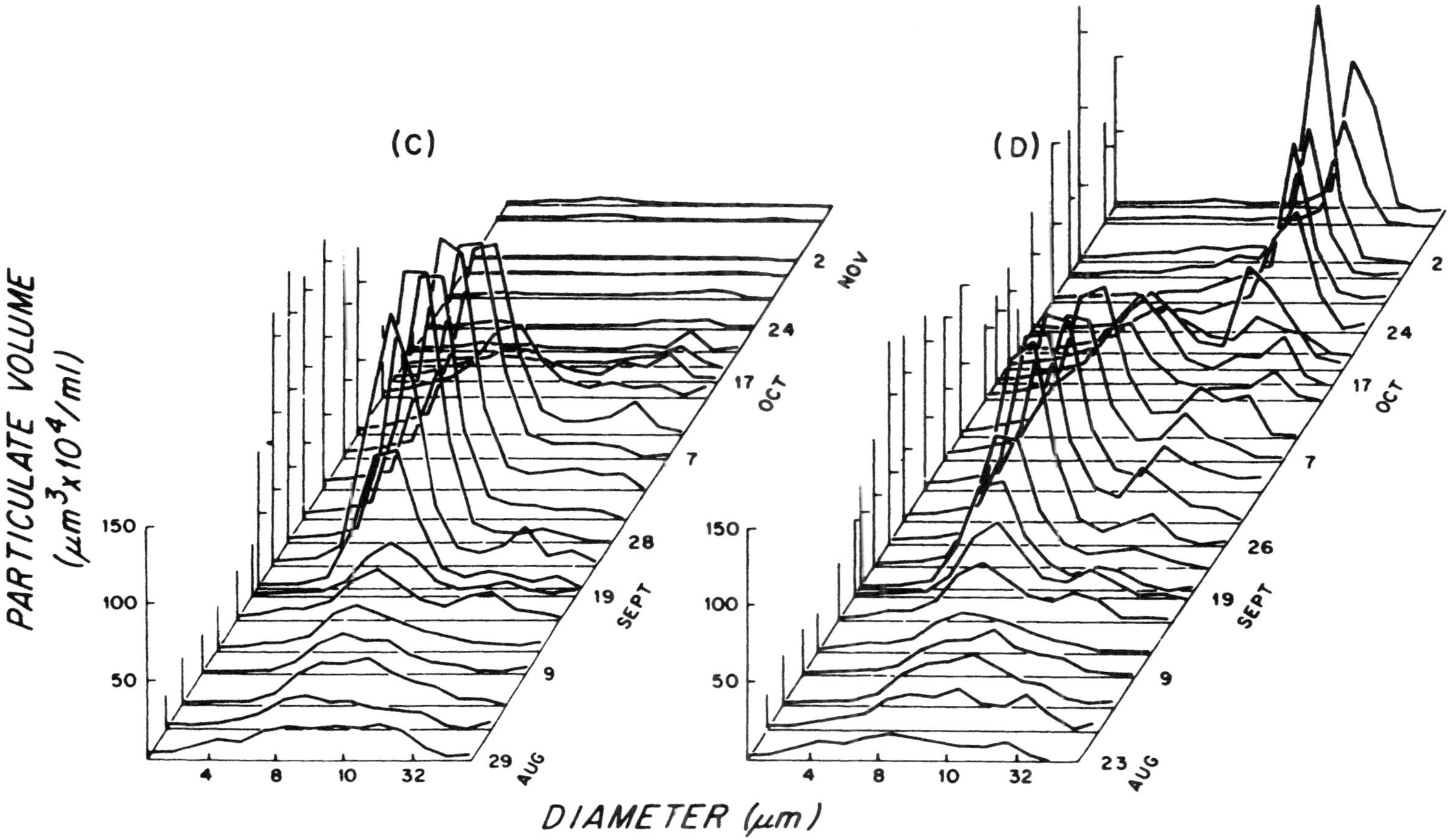

Figure 10. Particle size distribution at 5 m depth in enclosures C and D. From Steele and Gamble (1982).

nitrogen, followed by an increase of particulate material, and finally also ammonia in the bottom layer, reflected the flux of nitrogen very well (Fig. 11).

Hollibaugh et al. (1980) employed incubation experiments with ^{3}H-labelled amino acids in combination with large mesocosms, and found very high fluxes of free amino acids ranging from 0.09 to 2.42 mM m^{-3} d^{-1}. Amino acid degradation in deeper layers could account for 60% of the flux into the ammonium pool. The carbon content of released free amino acids reached up to 78% of primary production.

Even spiked phosphate (^{32}P) has been used in enclosure research in a pond (Henry, 1985). The flux of the relative radioactivity was followed through different size-classes. Smaller size-classes of zooplankton cycled more phosphorus than larger species. The results indicated that predation was most significant for nutrient regeneration.

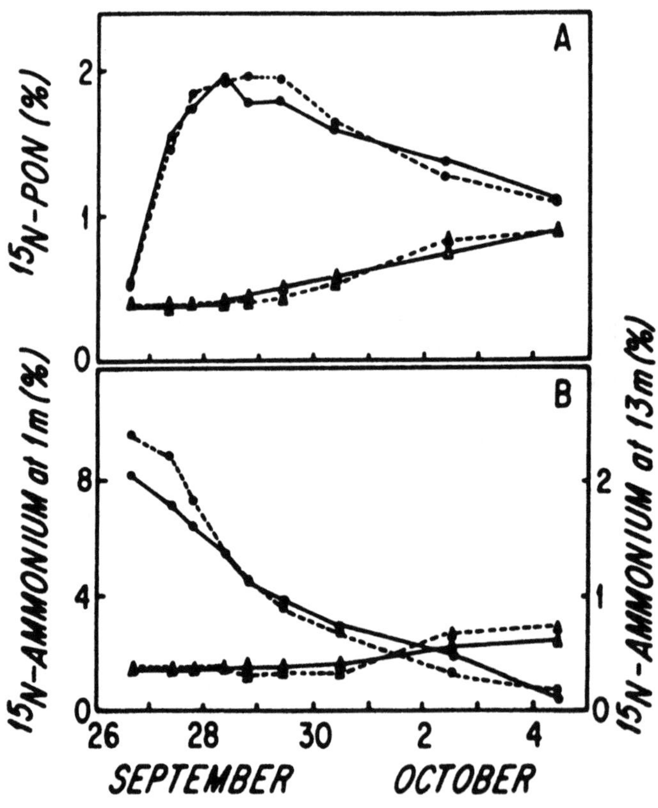

Figure 11. Changes with time of ^{15}N contents in (A) particulate organic nitrogen (PON) and (B) ammonium in CEEs in fall, 1977. Open circles, 1 m in CEE S (control); closed circles, 1 m in CEE R (5 μg Hg l^{-1} added); open triangles, 13 m in CEE S; closed triangles, 13 m in CEE R. From Koike et al (1982).

A MERL enclosure spiked with ^{14}C was used to study effects of a filter-feeding clam (*Mercenaria mercenaria*) on carbon cycling (Doering et al., 1986). Weekly measurements were sufficient in this benthic experiment. The limited profit of tracer experiments using the whole enclosure is evident by the uncontrolled loss of CO_2 which, in this study, was calculated from the deficit in the ^{14}C-budget.

RECOMMENDATIONS AND CONCLUSIONS

A combination of mesocosm research with laboratory experiments should be encouraged. Rate measurements of isolated organisms on a laboratory scale are helpful for the analysis of complex natural systems, even enclosed in mesocosms, and should be included more often. The use of isotope tracers in the enclosures is useful when the fate of the added compounds is followed by frequent measurements through individual compartments or compounds. Otherwise there will be no advantage over measuring concentrations of naturally dominating elements and compounds.

Generally, effective mesocosm research requires at least efforts equivalent to those of ocean-going research cruises. Therefore, the enclosure experiments should be used more intensively by cooperating specialists. Mesocosm studies provide the opportunity for testing laboratory-scale results under near field conditions. For future ecological research, trace analyses of such compounds as vitamins, inhibitors, and toxins are urgently needed.

Pollution research with mesocosms is often more easily funded, and the controls, which are necessary in any case, should be extended to baseline studies, causing only relatively small increases of overall expenses.

For effective use of the excellent possibilities of well-designed mesocosms in ecological research, it is necessary to establish research conditions which are stable for "the generation time of at least the third trophic level" of papers: i.e. reports, single process analysis, and network of ecosystem analysis. This could be achieved by cooperation of an interdisciplinary small group with experience in enclosure research, together with specialists covering all important aspects of ecological research during the field phases, in order to get the critical parameter-set for answering specific biological questions. These kind of projects need funding for a longer time period to ensure the publication of final synthesizing results.

LITERATURE CITED

Antia, N. J., C. D. McAllister, T. R. Parsons, K. Stephens, and J. D. H. Strickland. 1963. Further measurements of primary production using a large volume plastic sphere. *Limnol. Oceanogr.* **8**: 166-183.

Banse, K. 1982. Experimental marine ecosystem enclosures in a historical perspective. Pp. 11-24. **In**: G. D. Grice and M. R. Reeve [eds.], *Marine Mesocosms. Biological and Chemical Research in Experimental Ecosystems.* New York: Springer-Verlag.

Beers, J. R., M. R. Reeve, and G. D. Grice. 1977. Controlled ecosystem pollution experiment: Effect of mercury on enclosed water columns. IV. Zooplankton population dynamics and production. *Mar. Sci. Comm.* **3**: 355-394.

Bienfang, P. K. 1982. Phytoplankton sinking-rate dynamics in enclosed experimental ecosystems. Pp. 261-274. **In**: G. D. Grice and M. R. Reeve [eds.], *Marine Mesocosms. Biological and Chemical Research in Experimental Ecosystems.* New York: Springer-Verlag.

von Bröckel, K.. 1982. Sedimentation of phytoplankton cells within controlled experimental ecosystems following launching, and implications for further enclosure studies. Pp. 251-259. **In**: G. D. Grice and M. R. Reeve [eds.], *Marine Mesocosms. Biological and Chemical Research in Experimental Ecosystems.* New York: Springer-Verlag.

Brockmann, U. 1990. Enclosed plankton ecosystems in harbours, fjords and the North Sea - Release of dissolved organic substances. *Mar. Sci. Bull.* (In press).

Brockmann, U. H., K. Eberlein, G. Hentzschel, H. K. Schöne, D. Siebers, K. Wandschneider, and A. Weber. 1977a. Parallel plastic tank experiments with cultures of marine diatoms. *Helgol. Wiss. Meeresunters.* **30**: 201-216.

Brockmann, U. H., K. Eberlein, P. Hosumbek, H. Trageser, E. Maier-Reimer, H. K. Schöne, and H. J. Junge. 1977b. The development of a natural plankton population in an outdoor tank with nutrient poor seawater. I. Phytoplankton succession. *Mar. Biol.* **43**: 1-17.

Brockmann, U. H. and G. Hentzschel. 1983. Samplers for enclosed stratified water columns. *Mar. Ecol. Prog. Ser.* **14**: 107-109.

Brockmann, U. H., E. Dahl, and K. Eberlein. 1985. Nutrient dynamics during a *Gyrodinium aureolum* bloom. Pp. 239-244. **In**: Anderson, D. M., A. W. White and D. G. Baden [eds.], *Toxic Dinoflagellates.* New York: Elsevier.

Cadée, G. C. 1986. Organic carbon in the water column and its sedimentation, Fladen Ground (North Sea), May 1983. *Neth. J. Sea Res.* **20**: 347-358.

Dahl, E. and U. Brockmann. 1985. The growth of *Gyrodinium aureolum* Hulburt in *in situ* experimental bags. Pp. 233-238. **In**: D. M. Anderson, A. W. White, and D. G. Baden [eds.], *Toxic Dinoflagellates.* New York: Elsevier.

Dahl, E. and U. H. Brockmann. 1989. Does *Gyrodinium aureolum* Hulburt perform diurnal vertical migrations? Pp. 225-228. **In**: T. Okaichi, D. M. Anderson, and T. Nemoto [eds.], *Red Tides*. New York: Elsevier.

Davies, J. M. and P. J. leB. Williams. 1984. Verification of ^{14}C and O_2 derived primary organic production measurements using an enclosed ecosystem. *J. Plankton Res.* **6**: 457-474.

Davis, C. O. 1982. The importance of understanding phytoplankton life strategies in the design of enclosure experiments. Pp. 323-332. **In**: G. D. Grice and M. R. Reeve [eds.], *Marine Mesocosms. Biological and Chemical Research in Experimental Ecosystems.* New York: Springer-Verlag.

Doering, P. H., C. A. Oviatt, and J. R. Kelly. 1986. The effects of the filter-feeding clam *Mercenaria mercenaria* on carbon cycling in experimental marine mesocosms. *J. Mar. Res.* **44**: 839-861.

Eberlein, K. and U. H. Brockmann. 1986. Development of particulate and dissolved carbohydrates in parallel enclosure experiments with monocultures of *Thalassiosira rotula*. *Mar. Ecol. Prog. Ser.* **32**: 133-138.

Gamble, J. C. 1978. Copepod grazing during a declining spring phytoplankton bloom in the northern North Sea. *Mar. Biol.* **49**: 303-315.

Gamble, J. C. and J. M. Davies. 1982. Application of enclosures to the study of marine pelagic systems. Pp. 25-48. **In**: G. D. Grice and M. R. Reeve [eds.], *Marine Mesocosms. Biological and Chemical Research in Experimental Ecosystems.* New York: Springer-Verlag.

George, D. G. 1983. Interrelations between the vertical distribution of *Daphnia* and chlorophyll *a* in two large limnetic enclosures. *J. Plankton Res.* **5**: 457-475.

Goldman, J. C. and R. Mann. 1980. Temperature-influenced variations in speciation and chemical composition of marine phytoplankton in outdoor mass cultures. *J. Exp. Mar. Biol. Ecol.* **46**: 29-39.

Gust, G. 1977. Turbulence and waves inside flexible wall systems designed for biological studies. *Mar. Biol.* **42**: 47-53.

Hammer, K. D., U. H. Brockmann, and G. Kattner. 1981. Release of dissolved free amino acids during a bloom of *Thalassiosira rotula*. *Kieler Meeresforsch. Sonderh.* **5**: 101-109.

Hammer, K. D. and U. H. Brockmann. 1983. Rhythmic release of dissolved free amino acids from partly synchronized *Thalassiosira rotula* under nearly natural conditions. *Mar. Biol.* **74**: 305-312.

Hammer, K. D. and G. Kattner. 1986. Dissolved free amino acids in the marine environment: a carbon to nitrogen ratio shift during diatom blooms. *Mar. Ecol. Prog. Ser.* **31**: 35-45.

Harris, R. P., M. R. Reeve, G. D. Grice, G. T. Evans, V. R. Gibson, J. R. Beers, and B. K. Sullivan. 1982. Trophic interactions and production processes in natural zooplankton communities in enclosed water columns. Pp. 353-387. **In**: G. D. Grice and M. R. Reeve [eds.], *Marine Mesocosms. Biological and Chemical Research in Experimental Ecosystems.* New York: Springer-Verlag.

Harrison, P. J. and D. H. Turpin. 1982. The manipulation of physical, chemical and biological factors to select species from natural phytoplankton communities. Pp. 275-289. **In**: G. D. Grice and M. R. Reeve [eds.], *Marine Mesocosms. Biological and Chemical Research in Experimental Ecosystems.* New York: Springer-Verlag.

Harrison, W. G. and J. M. Davies. 1977. Nitrogen cycling in a marine planktonic food chain: Nitrogen fluxes through the principal components and the effects of adding copper. *Mar. Biol.* **43**: 299-306.

Henry, R. L. 1985. The impact of zooplankton size structure on phosphorus cycling in field enclosures. *Hydrobiologia* **120**: 3-9.

Hessen, D. O. and J. P. Nilssen. 1986. From phytoplankton to detritus and bacteria: effects of short-term nutrient and fish perturbations in a eutrophic lake. *Arch. Hydrobiol.* **105**: 273-284.

Hollibaugh, J. T., A. B. Carruthers, J. A. Fuhrman, and F. Azam. 1980. Cycling of organic nitrogen in marine plankton communities studied in enclosed water columns. *Mar. Biol.* **59**: 15-21.

Jahnke, J., U. H. Brockmann, L. Aletsee, and K. D. Hammer. 1983. Phytoplankton activity in enclosed and free marine ecosystems in a southern Norwegian fjord during spring 1979. *Mar. Ecol. Prog. Ser.* **14**: 19-28.

Kohata, K. and M. Watanabe 1986. Synchronous division and the pattern of diel vertical migration of *Heterosigma akashiwo* Hada (Raphidophyceae) in a laboratory culture tank. *J. Exp. Mar. Biol. Ecol.* **100**: 209-224.

Koike, I., A. Hattori, M. Takahashi, and J. J. Goering. 1982. The use of enclosed experimental ecosystems to study nitrogen dynamics in coastal waters. Pp. 291-303. **In**: G. D. Grice and M. R. Reeve [eds.], *Marine Mesocosms. Biological and Chemical Research in Experimental Ecosystems.* New York: Springer-Verlag.

Kuiper, J. 1977. Development of North Sea coastal plankton communities in separate plastic bags under identical conditions. *Mar. Biol.* **44**: 97-107.

Kuiper, J. 1981. Fate and effects of mercury in marine plankton communities in experimental enclosures. *Ecotoxicol. Environ. Saf.* **5**: 106-134.

Kuiper, J. and J. Gamble. 1988. Between test-tubes and North Sea mesocosms. Pp. 638-654. **In**: Salomons, W., B. Bayne, E. Duursma, and U. Förstner [eds.], *Pollution of the North Sea: An Assessment.* Berlin: Springer-Verlag.

McAllister, C. D., T. R. Parsons, K. Stephens, and J. D. H. Strickland. 1961. Measurements of primary production in coastal sea water using a large-volume plastic sphere. *Limnol. Oceanogr.* **6**: 237-258.

Oviatt, C. A., A. A. Keller, P. A. Sampou, and L. L. Beatty. 1986a. Patterns of productivity during eutrophication: a mesocosm experiment. *Mar. Ecol. Prog. Ser.* **28**: 69-80.

Oviatt, C. A., D. T. Rudnick, A. A. Keller, P. A. Sampou, and G. T. Almquist. 1986b. A comparison of system (O_2 and CO_2) and C-14 measurements of metabolism in estuarine mesocosms. *Mar. Ecol. Prog. Ser.* **28**: 57-67.

Parsons, T. R., P. J. Harrison, and R. Waters. 1978. An experimental simulation of changes in diatom and flagellate blooms. *J. Exp. Mar. Biol. Ecol.* **32**: 285-294.

Reeve, M. R., G. D. Grice, and R. P. Harris. 1982. The CEPEX approach and its implications for future studies in plankton ecology. Pp. 389-398. **In**: G. D. Grice and M. R. Reeve [eds.], *Marine Mesocosms. Biological and Chemical Research in Experimental Ecosystems.* New York: Springer-Verlag.

Salki, A., M. Turner, K. Patalas, J. Rudd, and D. Findlay. 1985. The influence of fish-zooplankton-phytoplankton interactions on the results of selenium toxicity experiments within large enclosures. *Can. J. Fish. Aquat. Sci.* **42**: 1132-1143.

Schindler, D. W. 1977. Evolution of phosphorus limitation in lakes. *Science* **195**: 260-262.

Smith, W., V. R. Gibson, and J. F. Grassle. 1982. Replication in controlled marine systems: Presenting the evidence. Pp. 217-225. **In**: G. D. Grice and M. R. Reeve [eds.], *Marine Mesocosms. Biological and Chemical Research in Experimental Ecosystems.* New York: Springer-Verlag.

Steele, J. H., D. M. Farmer, and E. W. Henderson. 1977. Circulation and temperature structure in large marine enclosures. *J. Fish. Res. Bd. Canada* **34**: 1095-1104.

Steele, J. H. and J. C. Gamble. 1982. Predator control in enclosures. Pp. 227-237. **In**: G. D. Grice and M. R. Reeve [eds.], *Marine Mesocosms. Biological and Chemical Research in Experimental Ecosystems.* New York: Springer-Verlag.

Takahashi, M., W. H. Thomas, D. L. R. Seibert, J. Beers, P. Koeller, and T. R. Parsons. 1975. The replication of biological events in enclosed water columns. *Arch. Hydrobiol.* **76**: 5-23.

Takahashi, M., I. Koike, K. Iseki, P. K. Bienfang, and A. Hattori. 1982. Phytoplankton species' responses to nutrient changes in experimental enclosures and coastal waters. Pp. 333-340. **In**: G. D. Grice and M. R. Reeve [eds.], *Marine Mesocosms. Biological and Chemical Research in Experimental Ecosystems.* New York: Springer-Verlag.

Williams, P. J. LeB. 1982. Microbial contribution to overall plankton community respiration - studies in enclosures. Pp. 305-321. **In**: G. D. Grice and M. R. Reeve [eds.], *Marine Mesocosms. Biological and Chemical Research in Experimental Ecosystems.* New York: Springer-Verlag.

6. BENTHIC MESOCOSMS: I. BASIC RESEARCH IN SOFT-BOTTOM BENTHIC MESOCOSMS

P. A. W. J. de Wilde

Abstract
Research in benthic mesocosms presents a relatively young and promising branch of science. Although primarily intended for ecotoxicological and other applied studies, there are also good examples of basic scientific results, significantly contributing to a better understanding of marine ecosystems. Innovations in the techniques and better research strategies, preferably in conjunction with field studies, are needed to improve the image of mesocosm research.

INTRODUCTION

This contribution will review benthic mesocosm facilities and the basic scientific results obtained from them. For pollution studies, see Pilson's contribution in this volume. The progress made in this field is demonstrated, conclusions are drawn, and the lines along which future developments may evolve are indicated.

Some earlier efforts dealing with mesocosm research are indicated, but emphasis is placed on existing experimental ecosystems that mimic soft-bottom environments and show a considerable degree of ecological realism. Those pelagic mesocosms to which benthic components were merely added are mentioned, but they have hardly contributed to fundamental studies. Freshwater mesocosms and enclosures are thought to fall beyond the scope of this contribution.

BENTHIC MESOCOSM FACILITIES

At present, major benthic mesocosm facilities are concentrated in the three centers described below.

1. Marine Ecosystem Research Laboratory (MERL) of the Graduate School of Oceanography, Narragansett, University of Rhode Island, U.S.A.

The MERL facilities, already initiated in 1975, now consist of 14 cylindrical tanks, each with a sediment container of 2.5 m^2 at the bottom. These outdoor mesocosms simulate the estuarine environment of Narragansett Bay (Santschi, 1985). Since its early start, MERL has been successfully operated to accommodate a wide range of applied and basic research programs including geochemical, ecological, and biological studies, and radioisotope experiments.

2. Marine Research Station at Solbergstrand of the Norwegian Institute for Water Research (NIVA), Oslo, Norway

Soft-bottom mesocosm facilities at Solbergstrand were realized in 1983 and consist of two large indoor basins, each 100 m^2, in which separate sediment sections are placed (Berge et al., 1986). In this way, subtidal bottoms from various places and depths of the Oslofjord and the North Sea became available for an experimental approach. The mesocosms were primarily intended for applied research. Many of the projects, however, have been performed jointly by NIVA and the University of Oslo, as this cooperation appeared to be beneficial to basic research.

3. Netherlands Institute for Sea Research (NIOZ) and Research Institute for Nature Management (RIN) in Texel, Netherlands

The mesocosms in Texel offer various possibilities for experimentation in soft-bottom environments. In 1975, indoor mudflat systems (20 m^2) were established (de Wilde and Kuipers, 1977) and used for a period of almost 10 years in basic studies (de Wilde, in press). In 1981, four pairs of outdoor Model Tidal Flats (MOTIFs), also with a surface area of 20 m^2, became available for applied research on the effects of oil spills and oil spill combating techniques on intertidal systems of the Wadden Sea. In 1987, in a modified set-up, four of the MOTIFs were connected to four Model Dump Sites (MODUS), in which configuration the effects of contaminated harbor sludge on coastal ecosystems were studied. Finally in 1988, two subtidal North Sea mesocosms, with a total surface area of 40 m^2, came into use. Largely following the Solbergstrand concept, either single sediment sections from the North Sea were transplanted or entire surface areas, composed from numerous boxcore samples, were accommodated.

In addition to these existing systems, several valuable initiatives dealing with former benthic mesocosm research will be mentioned here.

Firstly, the Oceanographic Institute in Kiel (F.R.G.) experimented with a large plankton tower in the Baltic Sea to follow the enclosed plankton community and to study the interaction with the natural bottom system underneath the water column (von Bodungen et al., 1976). Notwithstanding technical problems experienced in the efforts to couple the benthic and the pelagic systems, release of nutrients from the sediment could be documented and explained (Smetacek et al., 1976).

De Kock and Kuiper (1981) described a single enclosure experiment in which parts of natural mussel beds in the Dutch Wadden Sea were enclosed by wooden fences driven into the sediment. Although fully intended for ecotoxicological studies, the usefulness of similar structures for basic research is obvious. The same holds for an enclosure technique practiced by the Institute for Sea Research in Bremerhaven (F.R.G.) (Farke et al., 1984). The so-called Bremerhaven caissons are floating structures which can be lowered and consolidated on the intertidal mudflats of the Wadden Sea. Long-term pollution experiments were carried out in those enclosed areas. Unfortunately, the caisson experiments were stopped after some years of successful deployment.

In-situ benthic chambers enclosing subtidal communities in Scottish lochs were used by Davies et al. (1981), who realized coupling with the pelagic system by transfer of settled material derived from pelagic mesocosms, and by Eleftheriou et al. (1982), who added sewage to the chambers to study the effect on the benthic community. Bakke et al. (1982) studied the response of subtidal benthic communities to seawater extracts of diesel oil in *in-situ* deployed enclosures at 8 m depth in a Norwegian fjord.

Finally, in 1985-86, under the umbrella of a cooperative research program between the People's Republic of China and Canada, a number of mesocosm experiments were carried out (Proceedings IMEEES Symposium, Beijing, 1987, in press). Relevant benthic studies, in which trays containing contaminated sediments of different sources were attached underneath pelagic enclosures, all had an applied character.

SCIENTIFIC RESULTS FROM BENTHIC MESOCOSMS

1. MERL

MERL had several aims: to obtain insight into the biological interactions of complex systems; to investigate the effects of pollutants on ecosystems; and to study the biogeochemistry of selected pollutants in the estuarine environment (Pilson and Nixon, 1980).

From its early beginning, the chosen MERL concept proved to be useful. The multitrophic benthic communities, transplanted from the outside Narragansett Bay, survived well in the tanks and showed sufficient duration to allow long-term experimentation and observation. A number of tests showed that the inter- and intra-tank replicability of the MERL mesocosms was sufficient and adequate to the initiated experiments (Adler et al., 1980; Oviatt et al., 1980; Smith et al., 1982). From observations in the mesocosms and outside in the estuary, sufficient field-resemblance was concluded (Pilson and Nixon, 1980). Various ecosystem parameters were measured in the control tanks over a period of one year, and the majority were found to be typical of those observed in Narragansett Bay.

In a review article, Santschi (1985) has already documented and advocated the merits and results of ten years of MERL research, covering a wide range of applied and fundamental studies. Special attention was paid to the concentrations of nutrients in the mesocosms, from which it was concluded that the benthic compartment plays an important role in the availability of nutrients in the water column, and that the benthic flux of nutrients is partly a function of temperature (Pilson and Nixon, 1980; Oviatt et al., 1981a; Santschi et al., 1982). In general, a close benthic-pelagic coupling was found in the mesocosms, which apparently holds also for natural estuaries. The thickness of the benthic boundary film is of decisive significance for the flux rates of ions and molecules both from and toward the sediment, which is very important in understanding the mobility of contaminants. Quite unexpectedly, the rate of sediment resuspension proved to be largely controlled by the activity of the benthic fauna (Santschi et al., 1982). Benthic filter-feeding was found to be most important, enhancing removal rates of suspended particles and particle-associated contaminants (Santschi et al., 1983).

Inherent to mesocosm research is the ease with which complicated experiments, hardly possible in the field, are carried out in the MERL facilities. These include such things as the effects of a storm in the estuarine environment, controlled eutrophication, and the addition of hydrocarbons, trace metals and polluted sediments. The effects of a storm in the MERLs showed that the system returned to its prior state in a few days. Suddenly increased concentrations of pollutants in the water column returned quickly to their normal levels (Oviatt et al., 1981b). A series of nutrient additions to the MERLs showed divergent effects: at lower loads, production and biomass of all trophic levels were enhanced; at higher loads, however, a de-coupling of the primary from the secondary production processes occurred, which induced large oscillations in production, pH, and oxygen saturation (Nixon et al., 1984). Further nutrient and eutrophication experiments in the MERLs contributed to a better understanding of nitrification and denitrification processes in coastal ecosystems (Berounsky and Nixon, 1985; Seitzinger and Nixon, 1985), of the effect of eutrophication on benthic communities (Grassle et al., 1985), of the cycling of nutrients across a eutrophication gradient (Kelly et al., 1985), and of patterns of productivity (Oviatt et al., 1986). A few mesocosm studies were restricted to the species level: egg production in a copepod (Donaghay, 1985); carbon cycling affected by a clam (Doering et al., 1986); and reproduction of a polychaete (Levin, 1986).

Twelve years of MERL activities have now resulted in almost 200 publications, more than half of which are widely distributed in journals and books. About two-thirds of the experiments dealt with applied studies; the rest are of purely scientific interest. Santschi (1985) has stated correctly that the results and observations in the MERL mesocosms have significantly contributed to a better understanding of numerous basic processes in coastal and estuarine ecosystems. Discovery of many of the relationships pertaining to benthic-pelagic coupling in estuaries would have been more difficult or even impossible to make in the field.

2. Solbergstrand

The start of the sublittoral soft-bottom project at Solbergstrand dates back only five years. The first experiments aimed at finding the optimal sampling and transplantation techniques and how to keep benthic communities from depths down to 200 m in the mesocosms (Berge et al., 1986). From numerous observations and measurements over long periods, both in the field and in the mesocosms, a fair field-resemblance was concluded in terms of pore-water chemistry, bacterial production, sedimentation, animal recruitment and mortality, etc. Differences in sediment chemistry could be explained by compaction of the sediments during transport, by reduced diffusion coefficients, and by the reduced supply of organic matter. Dominant macrofauna species retained their original densities. Filter feeders, however, suffered from reduced sedimentation rates, but bacterial production and carbon content in the sediment were in the same range as measured in the field. A remarkable observation was made of a flourishing population of a rare, solitary bryozoan species in the mesocosms (Berge et al., 1985), pointing to the special conditions which may prevail in mesocosms. Even when most of the *in-situ* environmental conditions are sufficiently simulated in a mesocosm, differences in predation pressure, competition for food, etc., may favor certain organisms.

Other studies carried out dealt with the effects of organic enrichments of soft-bottom communities, the effects of ophiuroids on community structure, and bacterial production. Research on the effects of contaminated drill cuttings and mine tailings on the seabed is excluded from this paper.

A good example of the specific research potential of mesocosms is given by Warwick et al. (1986). They studied the abundance and community structure of meiofaunal organisms in the immediate vicinity of fecal mounds of a polychaete species. Interesting distribution patterns showed the importance of macrofauna-meiofauna interactions in this respect. Thus, whereas sampling in sublittoral benthic communities commonly occurs more or less at random or at best at areal scales of meters, mesocosm facilities permit detailed areal orientation in the benthic environment at the centimeter or even millimeter scale. This permits detailed observations on the fine structure of the seabed.

3. Texel

Benthic mesocosm research in the Netherlands has been partly reviewed by de Wilde (in press). For ten years, two intertidal mudflats at the Netherlands Institute for Sea Research were solely intended for basic studies. During the first year, both systems, simulating a densely populated lugworm habitat, developed in an almost identical way. From this replicability, lifetime, and field-resemblance, the usefulness of the chosen concept was concluded.

Later, one of the mesocosms was used in several geochemical and biological studies (Vosjan and Olanczuk-Neyman, 1977; Baumfalk, 1979; Boon and Haverkamp, 1979; de Wilde and Berghuis, 1979; Witte and de Wilde, 1979), contributing to the knowledge of the Wadden Sea ecosystem.

In the other mesocosm, which was left undisturbed for almost ten years, the evolution of the polychaete-dominated community was followed (de Wilde, in press). Numbers and biomass of *Nereis* increased dramatically at the cost of *Arenicola*. Total biomass stabilized at a high level. Most remarkable, however, was the extraordinarily high primary productivity when compared to the relatively low light input to the system. Most likely, algal strains had been selected that were adapted to the uniform, low-light conditions in the mesocosm.

From 1981 onward, mesocosm research on the effects of oil and harbor sludge contamination was carried out in Texel by a joint research group in which several institutes and the Ministry of Transport and Public Works participated. Notwithstanding the applied nature of the studies, an implicit basic component was still present. Thus in 1983, for 8 months, the complete set of eight mesocosms was available to investigate the replicability of the experimental ecosystems (Kuiper et al., 1984a). The majority of the physico-chemical parameters and the biota in the water column and the benthos were found to develop in a similar way. Also, ETS activity and meiofauna structure were comparable in the systems. Sometimes larger variations in the development of the macrofauna were observed, but these could be attributed to technical or logistic problems during the initial phase. Moreover, the development of the measured parameters resembled well those found in the natural mudflats. Further, as a rule, two untreated blanks were always present and equally observed in each set of applied experiments (Kuiper et al., 1984b; Scholten et al., 1987). This resulted in numerous basic observations on key processes of the intertidal mudflat system including larval settlement of macrofaunal organisms, wax and wane of "summer annuals", growth and production, mineralization, etc. Of special interest was the observation that macrofauna biomass in the model tidal flats was about equal, irrespective of the composition of faunal assemblages in the tanks (Dekker and van Moorsel, 1987).

In late 1987, the delivery of two subtidal North Sea mesocosms gave a new impulse to benthic research in Texel. Pilot studies now underway aim at the establishment of large coherent surface areas composed from smaller pieces of North Sea bottom. A second goal will be to investigate how to manage and sample these systems without major disturbances.

GENERAL REMARKS AND CONCLUSIONS

So far the present generation of experimental ecosystems has been realized in a rather pragmatic way. Design and construction of most mesocosms resulted from general knowledge of the selected type of ecosystem, the demands of the experiments envisaged, and the available funding. Of major importance is the establishment and long-lasting preservation of a realistic benthic community. With caisson-like and other enclosure techniques, in which parts of natural ecosystems are captured, the initial quality of the benthic community is self-evident. Scour, leakage and uncontrolled flushing, however, interfere with the objectives. Often weather conditions prevent *in-situ* facilities from longer lifetimes.

Land-based structures, therefore, seem preferable, but these need to be stocked with a benthic system. Except for unstable and heavily turbated sediments, such as are found in intertidal areas, that can be easily transferred in bulk to the mesocosm and matured afterwards by stocking the sediment with adult organisms, most subtidal bottoms need to be collected and transplanted with great care in order to preserve the natural orientation of the sediments. Modern box samplers, which take sediment sections of 0.25 m^2 or more, are most suited. Larger areas can be composed from larger numbers of cores.

To validate the mesocosm, at least part of the established benthic system must be kept for a longer period at *in-situ* conditions to compare the changes occurring in the system with field data.

Research in benthic mesocosms, because of their relatively small surface areas, requires special sampling methods. Bottom sampling in mesocosms can be carried out very precisely, and this largely accentuates the mesocosm approach.

Some general recommendations, aiming at proper sampling in benthic mesocosms, are given:

1. Lower the overlying watertable to facilitate manual operations.

2. Never enter the surface areas of the sediment by foot; instead install a gantry.

3. Introduce a system of fixed coordinates, which provide determinate sampling squares.

4. Avoid sampling of the marginal zones of the mesocosm to exclude artifacts from wall effects.

5. Adopt agreed sampling schedules in time and space.

6. Practice differentiation in sample size and frequency according to envisaged needs.

7. Fill in the empty core holes either with the finished sediment or by inserting an empty tube.

8. In contrast to sampling in natural systems, research in benthic mesocosms preferably relies on (almost) non-destructive sampling and measuring techniques such as: direct observation of life-traces at the surface; reading of (stereo) photographs, time-lag exposures, etc.; insertion and measuring with microsensors and probes (pH, O_2, S^-, current, etc.); and sampling with microcorers for meiofauna, microorganisms, sediment characteristics, etc.

Benthic mesocosms that mimic coastal or intertidal ecosystems do not raise special problems as to construction, control of ambient conditions, and research strategies. Subtidal systems, however, particularly those of greater depths, are increasingly difficult to handle. Although the effects of the diminished hydrostatic pressure in the experimental tanks seem to be absent or unknown, criteria for most other environmental factors are more prominent. Designs of realistic deep-sea mesocosms are still futuristic.

Mesocosm research presents a young branch of science, which is still in its crystallization phase. Notwithstanding that the use of marine mesocosms often is associated with applied studies, these many-sided facilities may provide a powerful research tool for solving oceanographic problems. Even when mesocosms are established to serve the needs of applied research, they may often provide significant results of general interest.

From the scientific results discussed earlier, it appears that experimental studies in mesocosms already cover a fairly wide field of oceanography, ranging from geochemical and integrated ecological studies to studies on the functioning of single organisms. Special attention should be given to the results obtained from the MERL mesocosms in studying benthic-pelagic coupling in estuarine environments. Quantitatively, the scientific results are still modest but, with respect to the apparent ecosystem reality, many of the results are convincing and easy to extrapolate to the field. From a statistical point of view, mesocosm results may have a scent of suspicion, often due to the restricted set-up of the experiments, lack of replicates or pseudo-replicates, lack of blanks, etc. Sometimes mesocosm research can be misleading, or even dangerous, when the experiments and results have not been carefully planned and interpreted. Mesocosms preferably have to be used in conjunction with, or as an extension of, field studies. They are never meant as real substitutes for the natural marine environment. Finally, it has to be said that the financial aspects of proper mesocosm studies are not so much smaller than those of field studies. Irrespective of the investments in technical facilities, both approaches require the same dedication and efforts of a multidisciplinary team of experts and technicians.

OUTLOOK

Mesocosms must be regarded now as valid tools in oceanographic research; facilities and techniques, however, need updating. Development of a new generation of benthic mesocosms, especially dealing with subtidal bottoms in shelf sea areas, is evident.

As both the variations between different benthic ecosystems and the aims of investigation are large, standardization in benthic mesocosm designs will be difficult to realize. Still, there is much in common: containment of the sediment; maintenance of the benthic community; and sampling, automation, and computerization of environmental data, etc.

Three lines of innovation in benthic mesocosm facilities can be foreseen. The first aims at the development of large high-technology mesocosms, amply provided with system-regulation technology, advanced data-acquisition techniques, and real-time data processing for physical and chemical as well as biological parameters. Such facilities most likely will be realized in a few land-based centers, specialized in benthic mesocosm research. A second line points to the development of *in-situ,* large-scale, pelago-benthic mesocosms, to provide an experimental system for the quantification of the energetic, particulate, and mineral fluxes between the benthic and pelagic marine environments. Finally, there may be a need for low cost (mobile) benthic systems, so called "box cosms", which aim at small experimental units for incubation of undisturbed benthic cores.

LITERATURE CITED

Adler, D., M. Amdurer, and P. Santschi. 1980. Metal tracers in two marine microcosms: Sensitivity to scale and configuration. Pp. 348-368. **In**: J. P. Giesy, Jr. [ed.], *Microcosms in Ecological Research*. DOE Symposium Series 52, CONF-781101. Springfield, VA: National Technical Information Service.

Bakke, T., T. Dale, and T. F. Thingstad. 1982. Structural and metabolic responses of a subtidal sediment community to water extracts of oil. *Neth. J. Sea Res.* **16**: 524-537.

Baumfalk, Y. A. 1979. Heterogeneous grain size distribution in tidal flat sediment caused by bioturbation activity of *Arenicola marina* (Polychaeta). *Neth. J. Sea Res.* **13**: 428-440.

Berge, J. A., H. B. Lainaas, and K. Sandoy. 1985. The solitary bryozoan, *Monobryozoon limicola* Franzén (Ctenostomata), a comparison of mesocosm and field samples from Oslofjorden, Norway. *Sarsia* **70**: 91-94.

Berge, J. A., M. Schaanning, T. Bakke, K. A. Sandoy, G. M. Skeie, and W. G. Ambrose. 1986. A soft-bottom sublittoral mesocosm by the Oslofjord: Description, performance and examples of application. *Ophelia* **26**: 37-54.

Berounsky, V. M. and S. W. Nixon. 1985. Eutrophication and the rate of net nitrification in a coastal marine ecosystem. *Estuarine Coastal Shelf Sci.* **20**: 775-781.

von Bodungen, B., K. von Bröckel, V. Smetacek, and B. Zeitzschel. 1976. The plankton tower. I. A structure to study water/sediment interactions in enclosed water columns. *Mar. Biol.* **34**: 369-372.

Boon, J. J. and J. Haverkamp. 1979. Pyrolysics mass spectrometry of a benthic marine ecosystem - The influence of *Arenicola marina* on the organic matter cycle. *Neth. J. Sea Res.* **13**: 457-478.

Davies, J. M., R. Hardy, and A. D. McIntyre. 1981. Environmental effects of North Sea oil operations. *Mar. Pollut. Bull.* **12**: 412-416.

Dekker, R. and G. W. N. M. van Moorsel. 1987. Effects of different oil doses, dispersant and dispersed oil on macrofauna in model tidal flat ecosystems. Pp. 117-131. **In**: J. Kuiper and W. J. van den Brink [eds.], *Fate and Effects of Oil in Marine Ecosystems*. Dordrecht: Nijhoff Publ.

Doering, P. H., C. A. Oviatt, and J. R. Kelly. 1986. The effects of the filter feeding clam *Mercenaria mercenaria* on carbon cycling in experimental marine mesocosms. *J. Mar. Res.* **44**: 839-861.

Donaghay, P. L. 1985. An experimental test of the relative significance of food quality and past feeding history to limitation of egg production of the estuarine copepod *Acartia tonsa*. *Arch. Hydrobiol.* **21**: 235-245.

Eleftheriou, A., D. C. Moore, D. J. Basford, and M. R. Robertson. 1982. Underwater experiments on the effects of sewage sludge on a marine ecosystem. *Neth. J. Sea Res.* **16**: 465-473.

Farke, H., M. Schulz-Baldes, K. Ohm, and S. A. Gerlach. 1984. Bremerhaven caisson for intertidal field studies. *Mar. Ecol. Prog. Ser.* **16**: 193-197.

Grassle, J. F., J. P. Grassle, L. S. Brown-Leger, R. F. Petrecca, and N. J. Copley. 1985. Subtidal macrobenthos of Narragansett Bay. Field and mesocosm studies of the effects of eutrophication and organic input on benthic populations. Pp. 421-434. **In**: J. S. Gray and M. E. Christiansen [eds.], *Marine Biology of Polar Regions and Effects of Stress on Marine Organisms.* New York: Wiley.

Kelly, J. R., V. M. Berounsky, S. W. Nixon, and C. A. Oviatt. 1985. Benthic-pelagic coupling and nutrient cycling across an experimental eutrophication gradient. *Mar. Ecol. Prog. Ser.* **26**: 207-219.

de Kock, W. C. and J. Kuiper. 1981. Possibilities for marine pollution research at the ecosystem level. *Chemosphere* **10**: 575-603.

Kuiper, J., R. Dekker, P. A. W. J. de Wilde, and W. J. Wolff. 1984a. De ontwikkeling van een model wadecosysteem ten behoeve van ecotoxicologisch onderzoek. TNO rapport R84/141: 1-79.

Kuiper, J., P. A. W. J. de Wilde, and W. J. Wolff. 1984b. Oil pollution experiment (OPEX). I. Fate of an oil mousse and effects on macrofauna in a model ecosystem representing a Wadden Sea tidal mudflat. Pp. 331-359. **In**: G. Persoone, E. Jaspers, and C. Claus [eds.], *Ecotoxicological Testing for the Marine Environment*. Vol. 2. Bredene, Belgium: State Univ. Ghent and Inst. Mar. Scient. Res.

Levin, L. A. 1986. Effects of enrichment on reproduction in the opportunistic polychaete *Streblospio benedicti* (Webster): A mesocosm study. *Biol. Bull.* **171**: 143-160.

Nixon, S. W., M. E. Q. Pilson, C. A. Oviatt, P. Donaghay, B. Sullivan, S. Seitzinger, D. Rudnick, and J. Frithsen. 1984. Eutrophication of a coastal marine ecosystem - an experimental study using the MERL microcosm. Pp. 105-135. **In**: M. J. R. Fasham [ed.], *Flows of Energy and Materials in Marine Ecosystems: Theory and Practice*. New York: Plenum.

Oviatt, C. A., H. Walker, and M. E. Q. Pilson. 1980. An exploratory analysis of microcosm and ecosystem behavior using multivariate techniques. *Mar. Ecol. Prog. Ser.* **2**: 179-191.

Oviatt, C. A., B. A. Buckley, and S. W. Nixon. 1981a. Annual phytoplankton metabolism in Narragansett Bay calculated from survey field measurements and microcosm observations. *Estuaries* **4**: 167-175.

Oviatt, C. A., C. D. Hunt, G. A. Vargo, and K. W. Kopchynski. 1981b. Simulation of a storm event in marine microcosms. *J. Mar. Res.* **39**: 605-626.

Oviatt, C. A., A. A. Keller, P. A. Sampou, and L. L. Beatty. 1986. Patterns of productivity during eutrophication: a mesocosm experiment. *Mar. Ecol. Prog. Ser.* **28**: 69-80.

Pilson, M. E. Q. and S. W. Nixon. 1980. Annual nutrient cycles in a marine microcosm. Pp. 753-778. **In**: J. P. Giesy, Jr. [ed.], *Microcosms in Ecological Research*. DOE Symposium Series 52, CONF-781101. Springfield, VA: National Technical Information Service.

Santschi, P. H. 1985. The MERL mesocosm approach for studying sediment-water interactions and ecotoxicology. *Environ. Tech. Lett.* **6**: 335-350.

Santschi, P. H., S. Carson, and Y. H. Li. 1982. Natural radionuclides as tracers for geochemical processes in MERL mesocosms and Narragansett Bay. Pp. 97-109. **In**: G. D. Grice and M. R. Reeve [eds.], *Marine Mesocosms. Biological and Chemical Research in Experimental Ecosystems.* New York: Springer-Verlag.

Santschi, P. H., D. Adler, and M. Amdurer. 1983. The fate of particles and particle-reactive trace metals in coastal waters: radioisotope studies in microcosms. Pp. 331-349. **In**: C. S. Wong, E. Boyle, K. W. Bruland, J. D. Burton, and E. D. Goldberg [eds.], *Trace Metals in Sea Water*. New York: Plenum Press.

Scholten, M., T. Bowmer, H. van het Groenewoud, G. van Moorsel, P. A. W. J. de Wilde, Chr. Brouwer, and N. Dankers. 1987. Effects of a light crude (F3) and of two selected oil combat methods in experimental tidal flat ecosystems. Final report oil pollution experiments (OPEX) 1986. TNO Rep. Delft, Netherlands, R87/348: 1-78.

Seitzinger, S. P. and S. W. Nixon. 1985. Eutrophication and the rate of denitrification and N_2O production in coastal marine sediments. *Limnol. Oceanogr.* **30**: 1332-1339.

Smetacek, V., B. von Bodungen, K. von Bröckel, and B. Zeitzschel. 1976. The plankton tower. II. Release of nutrients from sediments due to changes in the density of bottom water. *Mar. Biol.* **34**: 373-378.

Smith, W., V. R. Gibson, and J. F. Grassle. 1982. Replication in controlled marine systems: representing the evidence. Pp. 217-225. **In**: G. D. Grice and M. R. Reeve [eds.], *Marine Mesocosms. Biological and Chemical Research in Experimental Ecosystems.* New York: Springer-Verlag.

Vosjan, J. H. and K. M. Olanczuk-Neyman. 1977. Vertical distribution of mineralization processes in a tidal sediment. *Neth. J. Sea Res.* **11**: 14-23.

Warwick, R. M., J. M. Gee, J. A. Berge, and W. Ambrose. 1986. Effects of feeding activity of the polychaete *Streblosoma bairdi* (Malmgren) on meiofaunal abundance and community structure. *Sarsia* **71**: 11-16.

de Wilde, P. A. W. J. (In press) Benthic mesocosms in the Netherlands. Proceedings IMEEES Symposium Beijing, 1987.

de Wilde, P. A. W. J. and B. R. Kuipers. 1977. A large indoor tidal mud-flat ecosystem. *Helgol. Wiss. Meeresunters.* **30**: 334-342.

de Wilde, P. A. W. J. and E. M. Berghuis. 1979. Laboratory experiments on growth of juvenile lugworms, *Arenicola marina*. *Neth. J. Sea Res.* **13**: 487-502.

Witte, F. and P. A. W. J. de Wilde. 1979. On the ecological relation between *Nereis diversicolor* and juvenile *Arenicola marina*. *Neth. J. Sea Res.* **13**: 394-405.

7. BENTHIC MESOCOSMS: II. BASIC RESEARCH IN HARD-BOTTOM BENTHIC MESOCOSMS

Torgeir Bakke

Abstract

The performance of hard-bottom experimental ecosystems has been evaluated on the bases of replicability in space and time, reproducibility of natural environmental factors, goodness in imitation of the original community, and suitability for studies of interaction between populations. Only two comprehensive systems have been reported in the literature, both imitating the littoral zone: an artificial *Fucus vesiculosus* community at Karlskrona, Sweden, and a boreal rocky shore community at Solbergstrand, Norway. Both are based on an open seawater supply. In addition, public aquaria hard-bottom communities have been used for research purposes, and the design of a coral reef system has been included as an example. Subtidal hard-bottom mesocosms have only recently been constructed, but no results have yet been produced.

These systems have been established by transplantation alone, or in combination with natural recruitment. In temperate water regimes, the latter will take several years to produce communities with dynamic stability. It is believed that community establishment time may be shortened by emphasizing careful transplantation rather than water-mediated recruitment. This would also improve the replicability among units.

The present systems have shown that hard-bottom mesocosms can be established and run successfully for several years. They have so far proved valuable in pollution research, and they should also have potential for basic research, especially in questions where one wants to manipulate the environmental regime, or population sizes and structure. Their advantage over *in situ* experimentation lies in the possibility of generating certain water regimes, better overall community control, and better protection of installations against damage. Their disadvantage is poor experimental replication and deviation among units with time. They should therefore not be regarded as replicates in laboratory terms, but as reasonably similar communities imitating hard-bottom conditions sufficiently to prevent the inherent organisms from reacting abnormally.

INTRODUCTION

A search in the literature reveals few experimental ecosystems dealing with the hard-bottom benthos. In a way this paucity is surprising because hard-bottom communities, both in the littoral and sublittoral, should lend themselves easily to mesocosm experimentation. The physical structure is in essence two-dimensional; artificial substrates can be created by use of concrete or rocks; and because most organisms are sessile, they can be moved without much disturbance into mesocosms by transplantation of their substrate rocks.

The main reason why only a few systems have been established is probably that *in situ* experiments are also relatively easily performed in natural hard-bottom communities, and with far less expense laid down in establishment and maintenance than with a typical mesocosm. There is also a substantial literature on *in situ* hard-bottom manipulation, especially in the littoral zone (e.g. Dayton, 1975; Lein, 1980; Bonsdorff and Nelson, 1981). Such field experiments cover all degrees of enclosure from natural rock-pools to cages, fences, and artificial substrates. Because they are generally fairly open to the ambient environment, they should not be considered as experimental ecosystems or "cosms" in the terms set forward by the Working Group, and they will not be treated here.

For basic questions of community structure and function and species interactions, field experiments are clearly preferable to experimental ecosystems because there is no doubt about the realism in the former. Still, there are situations where experimentation in natural habitats is prohibitive for logistical or other reasons, and where experimental ecosystems could be useful alternatives. In enclosures, the populations are more easily manipulated and controlled, which is a prerequisite for studies of population interactions and density-dependent processes. Enclosed systems are also less subjected to the vagaries of climate. Still, the prime advantage of hard-bottom mesocosms over field experiments is the opportunity they give to manipulate the environmental regime, such as water quality, light, temperature, substrate features, etc.

PRESENT HARD-BOTTOM MESOCOSM FACILITIES

The most comprehensive systems reported are the estuarine *Fucus vesiculosus* community mesocosms of the Baltic Sea laboratory in Karlskrona, Sweden (Notini et al., 1977), and the marine rocky shore mesocosms at the Marine Research Station, Solbergstrand, eastern Norway (Bakke, 1986; Gray, 1987).

The Karlskrona System

The Karlskrona system consists of a series of outdoor circular pools (6.6 m^2 surface area, 4.2 m^3 volume) lined with disposable polyethylene sheet (Fig. 1). The pools are supplied with running sea water with about 50% replacement every 24 hours. The water is taken from 4 m depth. The systems do not include any artificial wave action or tidal fluctuation, but currents can be generated by use of propellers.

Establishment of the community relies completely on transplantation. The bottom of the pools is covered with a layer of sand upon which are placed rocks with their epigrowth of fauna and macroalgae. For the purpose of standardized tests, one-third of the bottom is covered with stones overgrown with bladder-wrack, *Fucus vesiculosus*, and about 1 m^2 with stones with green algae. In addition, the basins contain buckets with silt for soft-bottom research purposes, and motile crustaceans and fish are introduced. About 30 species of macroscopic organisms are established in each mesocosm. *F. vesiculosus* constitutes about 90% of the community biomass, which is also characteristic of Baltic Sea littoral communities. In general, the pool communities are given 1 month to stabilize before any experiments are performed.

The Solbergstrand System

The Solbergstrand mesocosms consist of four outdoor concrete basins, each 8 x 5 x 1.5 m in size and containing 25 m^3 of sea water at mid-tide (Fig. 2). A wave generator, consisting of two heavy-duty PVC pipelines in a steel frame, runs along the entire side of each basin. The wave generator moves up and down mechanically (18 strokes min^{-1}) creating regular waves running across the basin to the other side, where the main model community is established on a series of steps constructed to imitate a slanted shore.

The basins are supplied with running sea water from 1 m depth. The water enters the basins through a header tank, ensuring a stable flow rate of 10 m^3 h^{-1} per basin. Water exchange rate is normally in the range of 2 to 4 hours. From each basin, a subsurface outlet leads to a flexible hose with a stiff nozzle that is raised and lowered automatically to regulate the basin water level in a 12 h sinusoidal cycle following the tidal rhythm and nominal amplitude (30 cm) on the shore. During extreme temperatures, the water is taken at 13 m depth to prevent overheating or ice formation.

Community establishment has been done once, in October, 1979, by transplantation of rocks with algae and animals from the littoral zone, and by self-propagation from larvae and spores entering the basins with the water. Community development was allowed for a period of 3 years before any experiments were conducted. At the end of this period, the basin

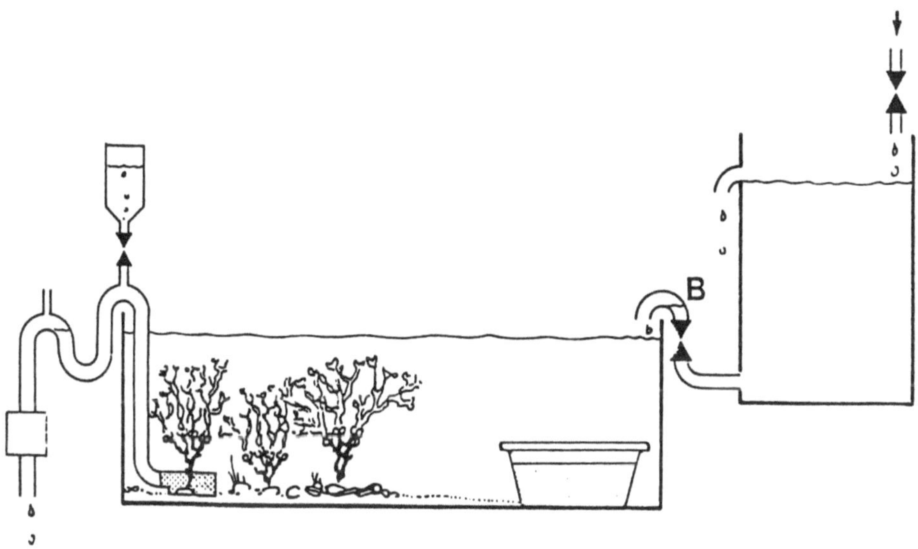

Figure 1. Diagram of the Karlskrona *Fucus vesiculosus* community mesocosms. After Notini et al. (1977).

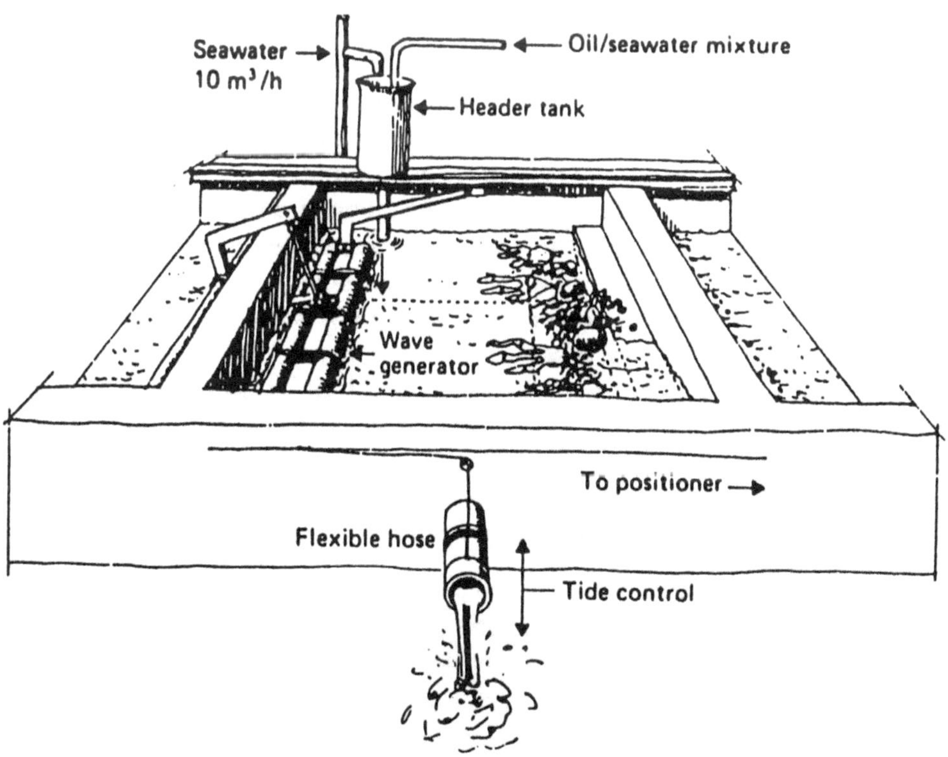

Figure 2. Diagram of one of the rocky shore mesocosms at Solbergstrand.

communities contained about 50 species of macroscopic algae and animals typical of medium-sheltered shores of the middle and outer Oslofjord region.

Other Facilities

In addition to the two Scandinavian systems designed for research, artificial hard-bottom communities are often constructed as elements of public exhibit aquaria. Some of these may be used for research as well, but the available literature contains little information of such activity. An example of design is the Caribbean coral-reef mesocosm established at the Smithsonian Institution, Washington, D.C., and described by Adey (1983). This system consists of a tank of 7 m^3 volume, in which a coral reef "base" has been established from dead carbonate material to simulate a typical reef configuration with a deep fore reef, a crest near the water surface, a back reef, and a sandy lagoon (Fig. 3). Water is supplied through a closed system, and currents and waves are sustained by pumps. The waves are created by shunting parts of the water through a container that turns over every 25 s, spilling the contents into the mesocosm.

The reef community is established by transplantation from coral reef habitats. The mesocosm contains about 35 algal species at any time, and 24 of the 45 species of stony corals, gorgonians, and anemones of the West Indian area are present. These species are reported to survive and grow for at least several months. The motile benthic fauna is relatively poor, but the fish fauna of the mesocosm comprises about 20 species. This is still only 10% of the number found in the natural habitat being imitated.

No report has been found of any mesocosm design that covers the sublittoral hard bottom explicitly, although some elements of the upper sublittoral zone are included in the current tidal systems. Thus the deepest steps of the Solbergstrand communities contain a typical subtidal *Fucus serratus/Laminaria saccharina* association. At present, a system with eight subtidal mesocosms (12 m^3 each) is established at Solbergstrand, based on transplantation of marine rock assemblages from a depth of 5 to 15 meters. The design will be similar to that of the Karlskrona system.

PRESENT RESEARCH IN HARD-BOTTOM MESOCOSMS

The scarcity of basic studies in hard-bottom mesocosms is even more pronounced than the scarcity of systems. Both the Karlskrona and the Solbergstrand systems have been established for pollution research purposes. The former is strongly standardized and marketed as an ecotoxicological test-apparatus in which effects of oil, dispersants, and pulp mill effluents, among others, have been tested. The latter has so far been used in a long-term study on effects of diesel oil (Bakke, 1986), and is at present applied in a study of

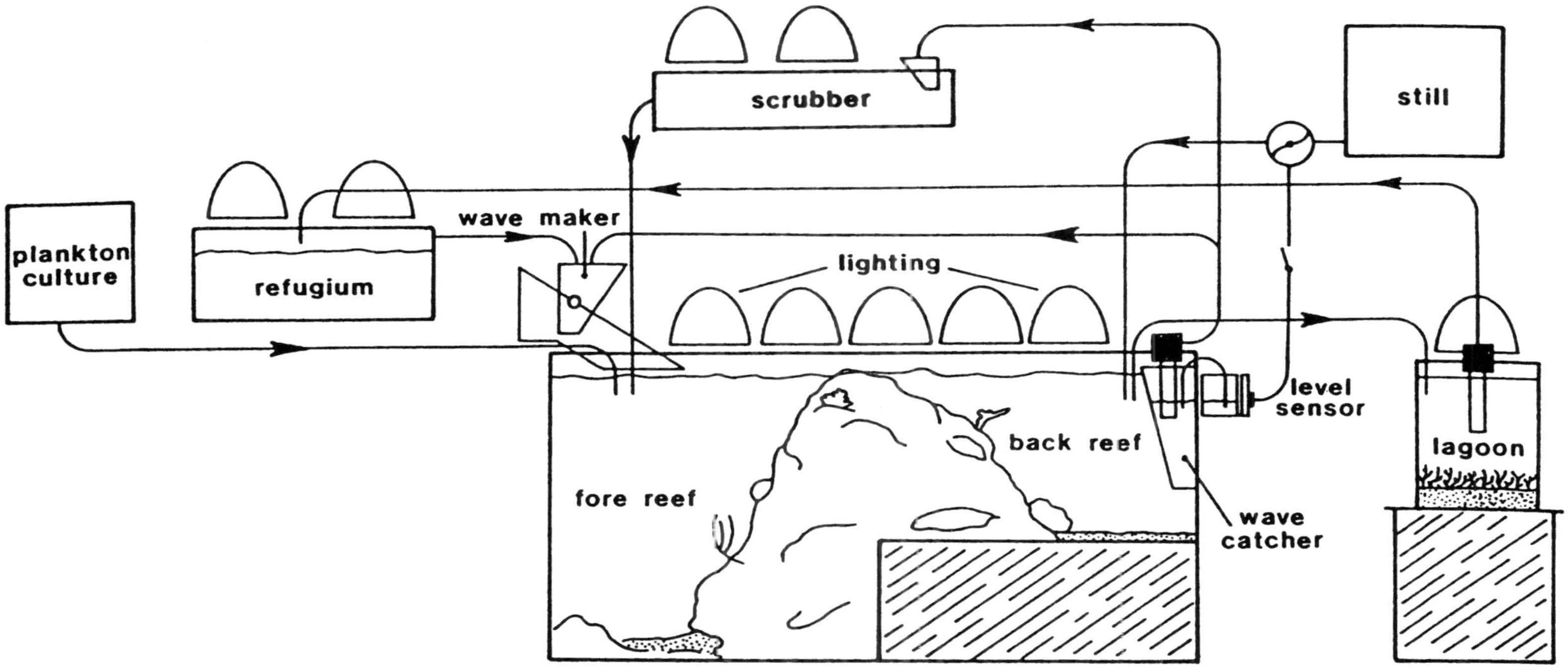

Figure 3. Diagram of the Smithsonian coral reef mesocosm. After Adey (1983).

settling success in *Ascophyllum nodosum* under various conditions of waves, immersion, and predation (Vadas, unpubl.)

These experiments show that the systems should have potential for basic research too. In general, the basic research elements included have been concentrated around mesocosm performance in comparison with the natural habitat in question.

The studies include differences in hydrographical regime, primarily whether extremes in temperature and nutrients are higher in the less-buffered mesocosm systems than in the comparable natural communities. In the Solbergstrand system, a considerable effort has been put into comparison of settlement and growth of sessile organisms on clean rock surfaces in mesocosms and in the field. Significant qualitative and quantitative differences have been recorded, partly explained by loss of recruits in the water system, partly by reduced water movement in the mesocosms. Diatom colonization studies have been performed at Karlskrona by use of glass substrates. Changes in population sizes and community structure in replicate mesocosms have been compared over 3 years at Solbergstrand. Study of community similarity over time is also part of the standard test parameters included in the Karlskrona studies. In the latter system, total community production and respiration are recorded by automatic logging of oxygen and pH. At Solbergstrand, similar effort has been concentrated on the primary rock community developed at intervals on clean granite tiles, and the effect of grazing on community metabolism has been included. Application of more than one control community in both systems has allowed comparison of community deviation with time, and the effects of this on population and individual performance.

Adey (1983) states that the coral-reef system has been used in manipulative experiments on animal behavior and on factors controlling primary production of the reef community, but no details have been given. It is thus obvious that, although the results have given valuable insight into how the current systems perform, the potential for manipulations useful for basic research purposes has not at all been utilized.

HOW DO PRESENT SYSTEMS PERFORM?

De Lafontaine and Leggett (1987) have suggested four criteria which should be fulfilled by an "ideal" enclosure system. It is clear that no single mesocosm design fulfills these requirements, but the present hard-bottom systems should be evaluated against them.

1. The design should allow good replication among enclosures in time and space

The Karlskrona system is established strictly by transplantation, and effort is made to stock the same total biomass of the major species in all enclosures. The stabilization period of 1 month is too short for any significant shift in community structure to occur before the experiments are started. Notini et al. (1977) reported that the similarity of the fauna associated with *Fucus vesiculosus* between two basins increased gradually during 2 months, based on a similarity index overemphasizing the rare species. Another index emphasizing the common species decreased slightly during the same period. This suggests that deviation represented change in densities rather than in species composition. Comparison of similarity in the fauna within the same basin over the 2 months revealed a tendency of deviation from the original faunal structure in one basin, but not in another.

At Solbergstrand, effort was also made to make the initial transplanted communities as similar as possible with respect to species composition. During the subsequent 3-year stabilization and recruitment phase, a dominance succession of keystone organisms was demonstrated. *Mytilus edulis* settled densely on walls, wave generators, and algae in all basins, both in 1979 and 1980, followed by the predators *Asterias rubens* and, slightly later, *Carcinus maenas*, both reducing the *M. edulis* population drastically during 1981-82. These predators were reduced after the summer of 1982, presumably due to lack of food and low winter temperatures. *Littorina littorea* and many of the macroalgae (*Cladophora rupestris, Fucus distichus* ssp. *edentatus, Laminaria saccharina, L. digitata* and *Phymatolithon lenormandii*) showed a gradual increase in density or cover during the establishment phase.

Although all basins showed the same overall pattern of succession, the intensity of population development was not the same in all. Hence after 3 years, at the start of the oil experiment, the replicate communities were quite different. The species composition and dominance pattern were the same in all communities, but keystone population densities and size structure differed, the former with a factor of up to five (for *L. littorea*). This was also reflected in the Bray-Curtis dissimilarity index (Clifford and Stephenson, 1975) based on the whole community, showing reasonable similarity between two of the communities and strong deviation in one (Bokn and Moy, unpubl.).

During the subsequent 3 years, the differences in population densities between the two control basins did not increase, and the difference in total community diversity (Shannon-Wiener index, Shannon and Weaver, 1963) did not diverge further (Bokn and Moy, unpubl.). Each control basin seemed to have a dynamic stability of its own, and somewhat different from the other.

Hence, good replication among communities of hard-bottom mesocosms can be achieved through careful transplantation, but such communities may lack the suite of less conspicuous species characterizing a natural community. Establishment through long-term recruitment can give reasonable replication in species composition, but one must expect significant deviation in population size and structure to develop among mesocosm units, which again have been shown to have impact on species interaction and density-dependent processes. Yet moderate, gradual, community deviation must be expected in rocky shore units, and it should be interpreted as an indication that no strong artificial forcing function is exerted on the community by the design or the researcher. Interestingly, the Solbergstrand experience has shown that a dynamic stability may be reached in enclosed rocky shore communities after some years and be sustained for at least 3 years, counteracting further deviation among the replicates. Such dynamic stability may or may not be reflected in a natural community, but it definitely eases long-term comparative experimentation. Since differences in recruitment are primarily due to factors that one should be able to minimize, such as differences in abundance of entering recruits and in physical properties of the mesocosm (e.g. substrate, currents, wave pattern), establishment of satisfactory replicates through combined transplantation and subsequent recruitment should be feasible.

2. The design should allow good reproducibility of the natural environment

The prime environmental factors to be reproduced in hard-bottom mesocosms are substrate, water regime, and light.

Characteristic substrate features such as texture, inclination, refugee areas (e.g. cracks and crevices) are easily imitated either by use of concrete or rocks, depending on the experimental scope. The steps of the Solbergstrand mesocosms may seem artificial, but in fact they only represent regular horizontal and vertical surfaces. By transplantation of rocks with epigrowth, the natural substrate is retained undisturbed.

For outdoor rocky shore mesocosms, light represents no problem. In the indoor Smithsonian coral-reef system, solar radiation is imitated by use of strong metal-halide vapor lamps sequentially switched on and off to simulate dawn, full daylight, and dusk. Also, the light period is extended to compensate for the lower light-energy input from the lamps compared to that of natural sunshine.

The main problem relates to water regime. In the Scandinavian systems, the water is renewed continuously from a natural shallow-water source and, provided sufficient turnover rate, temperature does not deviate significantly from the ambient. In the Karlskrona system, diel variation in oxygen and temperature was more pronounced than in the water of the natural littoral zone, even during September when air-water temperature difference is minimal (Notini et al., 1977). The same has been found in the Solbergstrand system only at mid-summer and mid-winter. During extreme summer heat, the basin water temperature has exceeded that of the shore water by a maximum of 2.6°C. In general, the long-term seasonal change in the temperature follows the surface of the fjord closely. The mean monthly temperature for the period 1980 to 1985 ranged from 0.6°C in March to 18.9°C in August, compared to 0.4°C and 18.3°C in the fjord (Bakke, 1986).

High water turnover also ensures constancy of water chemistry. In the Solbergstrand system, water turnover has to be shut off to be able to detect input/output differences in oxygen, pH, and nutrients due to community production and metabolism. In the Karlskrona system, this can be done without changing the water turnover rate. An interesting element in the coral reef design is a separate algal turf basin through which part of the water is shunted to increase oxygen, raise pH, and reduce excretory ammonia in order to balance the influences on water chemistry of the heterotrophically-dominated reef community.

The reason why hard-bottom communities exist at all is that water movement is too vigorous to allow sedimentation. The water movement is also an important regulation factor for the intensity of larval settlement and the relative abundance of species. Strong currents tend to favor certain assemblages, and must be imitated if such assemblages are to be supported for any length of time in a mesocosm. In the Solbergstrand mesocosms, currents are only generated by the velocity of the water entering the basins. This is sufficient to cause complete mixing of the basin water, but particle counts at the inlet and outlet have shown that the basins function as sediment traps. Furthermore, mussels and barnacles settle densely in the pipelines and header tanks leading to the basins, but far less in the basins themselves, although the recruits are present in sufficient abundance. Obviously the stimulus for settling created by water movement is lacking. Natural water movement may be difficult to achieve in a mesocosm due to the physical properties of the system, and neither of the present designs includes any advanced technology for generating currents across the substrate. Still, the technology required ought to be relatively simple.

For rocky shore mesocosms, waves and tidal fluctuation will have to be imitated, and the designs above are examples of how this may be achieved. A typical feature of these designs is that they are extremely regular, contrary to what is found on a natural shore. Such regularity creates sharply defined borders between zones of total desiccation, periodic air exposure, and total immersion, and in long-term studies this will result in sharp borders between sessile organisms with different demands on immersion period. This makes the whole community vulnerable to even slight maladjustment of the water level or wave pattern. In the Solbergstrand system, one of the basins was initially 6% (or 30 cm) wider than the others. This had only a slight influence on the wave pattern, but was sufficient to gradually change the distribution pattern of organisms on the steps. It was therefore necessary to move the wave generator and make the whole basin 30 cm narrower during the experiment.

A good mesocosm design should therefore also include technology to imitate natural and stochastic shifts in current speed and direction, as well as waves and tidal fluctuation (e.g. use of computer regulation of the water-movement generators).

3. The design should enable establishment of a representative community

This requirement leads directly to the question: what is a representative community? The structure of natural hard-bottom communities, especially in the littoral zone, is probably the best documented of any marine ecosystem, and both large-scale changes and mosaic recycling in time and space are normal features (e. g. see Hartnoll and Hawkins, 1980). A mesocosm experimenter can therefore claim, and probably correctly, that hard-bottom habitat areas exist for which his mesocosm is representative. Still, some basic requirements should be fulfilled.

Hard-bottom communities consist of a sessile and a motile element, and the generally dominating sessile element may again be divided into canopy, understory, and primary growth species. All of these basic groupings should be included, and are included in the current designs. The best way to establish representative hard-bottom experimental ecosystems is by enclosure of a natural habitat, because then all groupings are retained without disturbance. When basin communities are demanded, the two means to establish these are by transplantation and natural recruitment from larvae and propagules; present systems have shown that satisfactory hard-bottom communities can be established by these means. It is further believed that by careful and comprehensive transplantation of grown substrates and single populations, representative communities of both key species and the suite of less conspicuous species present on hard bottoms can be established in mesocosms without relying on longer-term recruitment.

4. The enclosure should allow observation of predator-prey relations at natural densities and ratios

Due to the relative ease in determining natural population densities and structure on hard bottoms, the information necessary to create good community imitations can be obtained. The possibility of studying the influence of top ranked vertebrate predators is, however, limited. In most cases, these are species of larger fish or birds, which normally cover large areas for feeding, and none of the present systems are large enough for even single individuals to be included for any length of time without unnatural overharvesting of the prey populations. Larger fish have therefore been excluded both in the Karlskrona and Solbergstrand systems, although smaller fish are part of the communities.

When first established, the sessile and two-dimensional nature of the hard-bottom mesocosm community makes observation of predator/prey and grazer/algal growth relations much easier than in pelagic and soft-bottom systems, especially since the observation generally can be made without disturbing the community. Such interaction studies are not reported from the Karlskrona experiments, but at Solbergstrand the influence of grazers on the pioneer growth community has been one of the main study objects (Pedersen, 1987). Also, amplification of oil pollution mortality in mussels through effects of predation have been recorded (Bakke, 1986), and effects of predatory crabs on periwinkle recruitment has been discussed (Lystad and Moe, 1985). Controlled manipulation of predator and prey densities should also be easily performed in hard-bottom mesocosms, but this possibility has not been exploited.

POTENTIALS AND FUTURE DEVELOPMENT

Hard-bottom model ecosystems have proven to be valuable experimental tools in pollution research, and they are the only alternatives to laboratory experiments if a chronic discharge into coastal water is to be imitated. Important ecological functions such as competition, predation, recruitment, and seasonality can be retained in basin communities for several years, and by careful management and advanced technical design, one should be able to copy the physical environment to satisfaction. Such mesocosms should therefore also be valuable in basic ecological research. They are superior to field experiments only for special purposes, because most of the sampling and automated techniques may in principle be applied to equally sheltered shores, and automatic recording of environmental conditions can also be arranged *in situ* if necessary.

Still, there are some clear logistical advantages with mesocosms. Experimental installations such as settling panels, cages, electrodes, etc. are better protected against accidental or deliberate damage, and sampling is very well controlled. Although a benthic mesocosm most often is connected to the outside world through the water supply, there is no real exchange of individuals with the surroundings, as in field experiments. This gives excellent control of population sizes. Also, dead individuals with hard structures are usually retained within the mesocosms and can be recorded.

The disadvantages of poor experimental replication and deviation of parallel communities with time exist in mesocosms as well as in field experiments, but are more manageable in the former. Even well-designed mesocosms should not be regarded as replicates in laboratory terms, but as a series of reasonably similar and partly independent communities imitating hard-bottom conditions sufficiently to prevent the inherent organisms from reacting abnormally in the experimental situation.

The experience from the present hard-bottom mesocosm designs suggests several improvements to be considered in future development. Better imitation of fluctuation and stochastic events in the physical regime of currents, waves, and tide is important, and may be obtained through more flexible technology coupled with computer regulation. Community-establishment time should be shortened by emphasizing careful transplantation more than water-mediated recruitment. This will also improve structural similarity in replicate communities. Efforts in enclosure techniques for littoral and sublittoral hard-bottom communities should also be pursued to obtain advanced *in situ* mesocosms, which would combine the realism of field experiments with the mesocosm possibility of managing the enclosed water.

LITERATURE CITED

Adey, W. H. 1983. The microcosm: a new tool for reef research. *Coral Reefs* **1**: 193-201.

Bakke, T. 1986. Experimental long term oil pollution in a boreal rocky shore environment. *Proceedings 9th AMOP Technical Seminar*, Canada, 1986. ISBN 0-662-14812-6.

Bonsdorff, E. and W. G. Nelson. 1981. Fate and effects of Ekofisk crude oil in the littoral of a Norwegian fjord. *Sarsia* **66**: 231-240.

Clifford, H. T. and W. Stephenson. 1975. *An Introduction to Numerical Classification*. New York: Academic Press. 229 pp.

Dayton, P. K. 1975. Experimental evaluation of ecological dominance in a rocky intertidal algal community. *Ecol. Monogr.* **45**: 137-159.

Gray, J. S. 1987. Oil pollution studies of the Solbergstrand mesocosms. *Phil. Trans. R. Soc. Lond. B* **316**: 641-654.

Hartnoll, R. G. and S. J. Hawkins. 1980. Monitoring rocky shore communities: a critical look at spatial and temporal variation. *Helgol. Wiss. Meeresunters.* **33**: 484-494.

de Lafontaine, Y. and W. C. Leggett. 1987. Evaluation of *in situ* enclosures for larval fish studies. *Can. J. Fish. Aquat. Sci.* **44**: 54-65.

Lein, T. E. 1980. The effects of *Littorina littorea* L. (Gastropoda) grazing on littoral green algae in the inner Oslofjord, Norway. *Sarsia* **65**: 87-92.

Lystad, E. and K. Moe. 1985. Comparison of population dynamics, growth and reproductive cycle in *Littorina littorea* (L.) at four artificial and one natural locality at Solbergstrand, Oslofjorden, where two of the artificial localities were exposed to hydrocarbons. University of Oslo: Unpubl. thesis for the Cand. Real. Degree. 198 pp. [In Norwegian.]

Notini, M., B. Nagell, A. Hagström, and O. Grahn. 1977. An outdoor model simulating a Baltic Sea littoral ecosystem. *Oikos* **28**: 2-9.

Pedersen, A. 1987. Community metabolism on rocky shore assemblages in a mesocosm: A. Fluctuations in production, respiration, chlorophyll *a* content and C:N ratios of grazed and non-grazed assemblages. *Hydrobiologia* **151/152**: 267-275.

Shannon, C. E. and W. Weaver. 1963. *The Mathematical Theory of Communication.* Urbana: University of Illinois Press. 118 pp.

8. SPECIFIC APPLICATION OF MESO- AND MACROCOSMS FOR SOLVING PROBLEMS IN FISHERIES RESEARCH

Victor Øiestad

Abstract

The great complexity of open-sea systems has made it necessary to study fish population dynamics, on a small scale, in the laboratory. These indoor studies have given a far better understanding of early life histories of fish species. However, when the results are applied in interpretation of open-sea events, the large gap in structure and scale makes this use of results dubious.

The need for transitional studies to bridge this gap has been suggested since the mid-1970s and increasing numbers of such studies have been carried out, mainly with commercial North Atlantic species. The main results from these studies have questioned the significance of some hypotheses set forward from laboratory observations. These include the hypotheses:

-that larval survival and growth is not dependent upon densities of food organisms in the order of 100 to 1,000 per liter, but rather in the order of 1 to 100;

-that predation might be responsible for a far larger fraction of larval mortality than earlier suggested; and

-that fatal starvation in the sea might be more a question of growth below a species-specific growth barrier for survival rather than larvae brought to a point of irreversible starvation as observed in traditional starvation studies.

Transitional studies are diversifying in the direction of small units (mesocosms) in large number and increasing use of larger systems (macrocosms). A further development should profit from integrated laboratory, meso- and macrocosm studies to deal with more specific and complex questions in early fish population dynamics. This strategy permits studies including post-metamorphosed stages. Larger programs within this field should utilize international coordination and participation, including a number of scientific disciplines.

INTRODUCTION

During the colloquium entitled "Larval Fish Mortality Studies and Their Relation to Fishery Research", held in La Jolla, California, in January, 1975, transition studies were recommended as a particularly promising method for larval research (Hunter, 1976). At the same time, a new generation of enclosure studies was initiated in Norway (Øiestad et al., 1976), more than a generation after earlier studies in the 1880s and 1930s (Rognerud, 1887; Rollefsen, 1946). These new studies were scaled-up further in 1980 and are still running.

The Norwegian research is part of extensive international activity in this field and covers only some types of enclosures in use. The other main centers for enclosure studies of fish larvae seem to be Scotland and the east coast of Canada.

Discussion of the "ideal" enclosure is still important, as the method is young and needs to develop further. De Lafontaine and Leggett (1987) suggested that "the 'ideal' enclosure should allow (1) good replication among enclosures, (2) good reproducibility of the natural environment, (3) the ability to establish a representative community, and (4) observation of predator-prey relations at natural densities and rations." Obviously it is easy to agree on these points, but as soon as a particular enclosure is evaluated against these prerequisites, opinion will differ. Most likely, any design will have some shortcomings. As a result, there might be a growing understanding of the need to combine different types of enclosures to give this tool a broader application and permit a wider range of items to be studied.

Until now, enclosure studies have mainly involved a few specialists. Interest from those responsible for stock management has so far been close to zero.

There is a strong need for an analysis of enclosure results obtained so far and of the potential use of the method, with particular attention to multispecies relations, which will be of increasing interest in fishery biology. The ultimate aim should be to integrate enclosure studies as a standard instrument in fishery research.

DESCRIPTION OF METHODS

Different types of enclosure studies have been the subject of several recent reviews, most recently by de Lafontaine and Leggett (1987), referred to with minor modifications in Table 1. It is convenient to distinguish between enclosures according to volume. Enclosures larger than 10,000 m^3 have elsewhere been called macrocosms, whereas those down to a volume of 1 m^3 have been called mesocosms.

Table 1.

Summary of enclosure types and uses for larval fish study. Modified after de Lafontaine and Leggett (1987).

Enclosure Description					
Wall Material	Z: depth (m) D: diameter (m) V: volume (m^3)	Duration (d)	Species	Study Objectives	References
Nonporous bags					
Black polyethylene	Z: 2.5 D: 0.95 V: 1.8	10-32 26 18-42	Herring Turbot Capelin	Growth, survival Growth Survival	Øiestad & Moksness, 1981 Rosenberg & Haugen, 1982 Øiestad, 1985
Transparent PVC bags (Loch Ewe)	Z: 19.5 D: 4.75 V: 310	60-80 86 86 56	Herring Herring Herring Cod	Growth, feeding Lipids content Lipids content Growth, feeding	Gamble et al., 1985 Fraser et al., 1987 Geffen, 1982 Gamble & Houde, 1984
Transparent polyethylene (CEPEX)	Z: 23.5 D: 10 V: 1300	31 45	Herring Salmon	Growth, feeding Growth	Houde & Berkeley, 1982 Koeller & Parsons, 1977
Porous bags					
White polyester	Z: 3.0 D: 2.0 V: 6.0	76	Herring	Survival	Schnack, 1981

White dacron	Z: 5.0 D: 1.0 V: 3.2	7-10	Capelin	Growth, survival	Leggett, 1986
Nitex, 153-μm mesh	Z: 5.0 D: 1.0 V: 3.2	3	Capelin	Predation mortality	Frank & Leggett, 1982
Nitex, 505-μm mesh	Z: 6.0 D: 1.7	14	Winter flounder	Survival	Laurence et al., 1979
Green knotless mesh, 9 & 19 mm	Z: 12.0 D: 3.1 V: 90	41	Salmon	Feeding	English, 1983
Basins					
Norwegian concrete basin	Z: 4.5 V: 4400	130-180 73 89 15-20 100	Cod Herring Herring Cod Capelin & herring	Growth, survival Growth, survival Otolith growth Condition factor Multispecies interaction	Ellertsen et al., 1981 Øiestad & Moksness, 1981 Geffen, 1982 Øiestad, 1984 Øiestad, 1985
Same as above	Z: 5.0 V: 2200	11 127	Turbot Capelin	Growth, survival Growth, survival	Rosenberg & Haugen, 1982 Moksness, 1982
Ponds					
Hyltro	Z: 6.0 V: 60000	180	Cod Cod	Growth, survival Intraspecies interaction	Øiestad et al., 1985b Øiestad et al., 1985a

One of the main objections to macrocosm studies is that replication is almost impossible for logistic and economic reasons. However, a macrocosm study should be considered more like a field study, where there is also no replication. The observer witnesses a dynamic process and, by sampling and monitoring, the significance of some specific processes can be examined. In a normal field situation the observer is in a rather unnerving situation, as the turbulence and movement of organisms do not permit him to follow the dynamic processes undisturbed. This problem is reduced in a macrocosm, as all processes take place within an enclosed water volume. Furthermore, the observer is able to control some of the conditions: the timing of larval release; the age of the larvae; the density of larvae; and, to some extent, the type and quantities of predators in the system as, in many macrocosms, water is filtered when introduced and the systems are treated with rotenone in advance.

After larval release, the observer continues monitoring larval survival, growth, and growth pattern in the population or cohort, the presence of predators and competitors, zooplankton composition and density, larval distribution and behavior, and a number of abiotic factors (Fig. 1). Furthermore, the population or cohort can be studied beyond metamorphosis in a macrocosm. Normally a collapse of the zooplankton population can be observed. Other investigations can be attached to this main stream of events. Some examples are:

(1) repeated release of the same species with a time-lag making identification of each cohort possible (Fig. 2), and then observation of the same phenomena under changed biotic and abiotic conditions;

(2) simultaneous release of other species, or delayed release for similar purposes; and

(3) repeated delayed releases, making it possible to identify if and when the main species starts to prey on the species released later, and the significance of the size of the predator species. The following year, the same macrocosm might be used to study the same species, with the main and delayed released species reversed. Thus it should be possible to mimic situations that might occur in the open sea, where a number of species coexist and where this type of situation often will occur.

Most macrocosm studies have been undertaken in Norway. Some major observations, as a further illustration of the method, will be dealt with later.

Mesocosm studies have been undertaken in a number of countries. In contrast to macrocosm studies, replication is the rule. In these studies, larval densities normally are higher ($>100\ m^{-3}$) than in macrocosm studies (2 to 50 m^{-3}), and the zooplankton community (in most cases at an early stage of the study) will not be able to supply food organisms at a rate equal to the grazing rate. The result is depletion of zooplankton and reduced larval growth rate

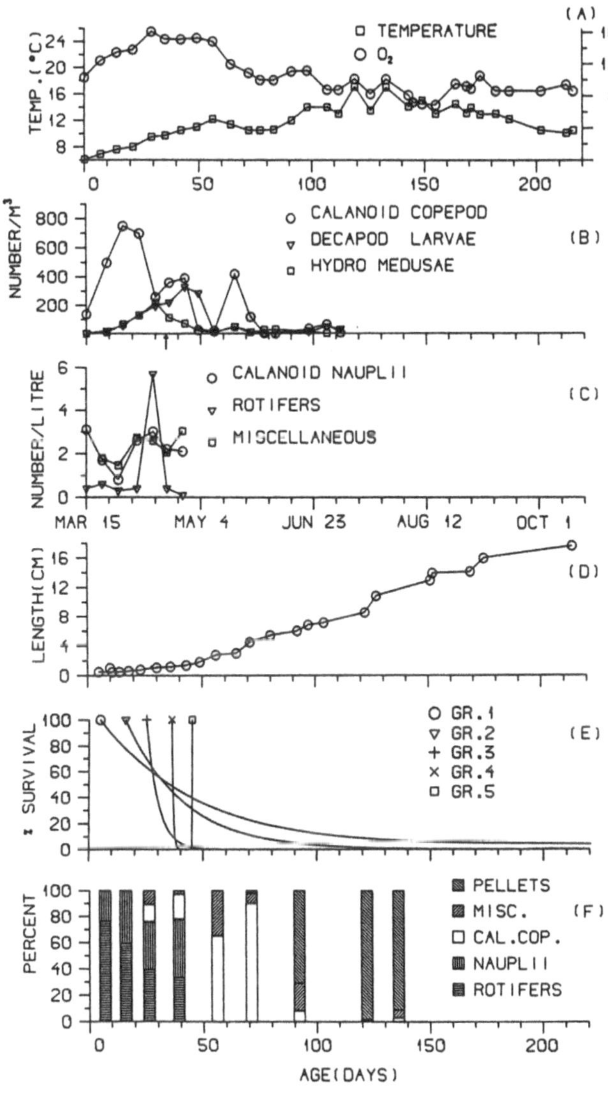

Figure 1. Conditions in the culture pond and biology of stocked Atlantic cod, 1983. Days are numbered from March 15, when the first group of Atlantic cod larvae hatched; the first stocking occurred on March 20 (day 5). A, Water temperature and oxygen saturation. B, Mean densities of calanoid copepods, decapod larvae, and hydromedusae. The arrow at day 35 indicates metamorphosis of the first group of Atlantic cod larvae. C, Mean densities of food organisms for first-feeding Atlantic cod larvae. D, Mean length of Atlantic cod larvae. E, Percent survival of five groups of Atlantic cod larvae; each group was stocked 5 days posthatch. F, Stomach contents of Atlantic cod larvae. Through day 80, data are frequencies of occurrence; thereafter, the volumetric fraction consisting of pellets was first estimated, then percent occurrence of organisms was calculated for the remaining fraction. From Øiestad et al. (1985b).

compared to most macrocosm studies. This drawback can partly be solved by a steady supply of zooplankton, but in many cases that will reduce the value of the study. Another solution is to have part of the bag as a fine-mesh net, permitting zooplankton to enter the system (Laurence et al., 1979; de Lafontaine and Leggett, 1987). The wall effect is more pronounced in mesocosms, but this is often overcome by producing the enclosures from a black material. This gives a more even distribution of larvae and zooplankton, but also gives the larvae a reduced light intensity (Fuiman and Gamble, 1988). This might give the larvae an illusion of living in deeper water than what they really experience. In most cases, these studies have to be terminated before, or a short time after, metamorphosis (de Lafontaine and Leggett, 1987). Sampling of large numbers of larvae is needed but might be undesirable at a late stage, as that will reduce the number of larvae left and may disturb the study.

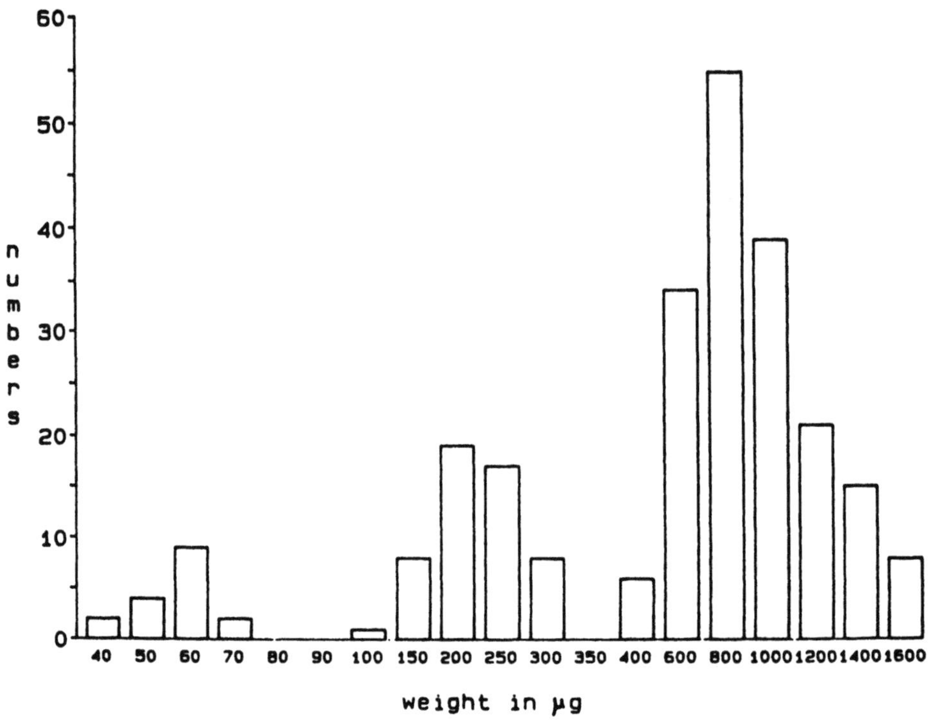

Figure 2. Dry weight distribution of Atlantic cod larvae in the 1983 Hyltro pond experiment with three cohorts present. Age from hatching for each of the cohorts is 30 days, 19 days, and 10 days respectively. Each cohort is easy to identify, and the largest larvae from the first cohort are about to metamorphose. From Øiestad et al. (1985a).

Mesocosm studies can provide an opportunity to observe most of the phenomena described in macrocosms, but repeated releases of larvae usually are of reduced value. Such a study might be done for a short duration, with terminations made as a time series (day 1, day 2, and so on). Multispecies studies have also been carried out (Gamble, 1984).

Mesocosm studies can be more powerful if they are linked to a macrocosm study (Øiestad and Moksness, 1981). As almost all macrocosms in use are in Norway, this combined method has not been fully exploited. Mesocosm studies can be linked with a main macrocosm study to elucidate special phenomena that mesocosm studies are particularly suited for. The effect on a larval cohort of food levels below those observed in the macrocosm, and the effect of shifting food supply, are two potential areas. Mesocosms might easily be supplied with filtered water (declining food level) and with sea water containing high concentrations of zooplankton; equipment for the last purpose is in frequent use in Norway (Fig. 3).

Special abiotic phenomena might also be mimicked in mesocosms as, for example, sea water supersaturated with nitrogen gas. Supersaturated sea water can be added to mesocosms at sublethal as well as lethal levels. The effect of vertical turbulence, which might be a major hazard at least for larvae with swimbladders, might be more difficult to mimic. This is one of many phenomena where laboratory studies should be included (in this case, by use of a pressure tank).

Houde (1985) strongly recommended comparison of results in different types of rearing systems, including laboratory-scale tanks. These studies should be carried out synoptically under similar environmental conditions.

If we imagine a study including all three levels, it is likely that no single research institute would have the personnel capacity of running an integrated experiment. The output should be far better if a number of institutes decided to run such an experiment together, in an appropriate place with access to all three types of facilities: macrocosms, mesocosms, and laboratory.

RESULTS

Instead of making brief comments on a large number of phenomena studied by use of meso- and macrocosms, I will concentrate on two phenomena which I consider to be essential in fish population dynamics and where transitional studies have contributed significantly to improve understanding: predation and growth barriers.

First, I will deal with what has been called the "growth barrier". Obviously there is an upper limit for specific growth rate (SGR) within a species, dictated primarily by intrinsic factors. In a number of enclosure studies, very high SGR values have been observed. Specific analysis of these

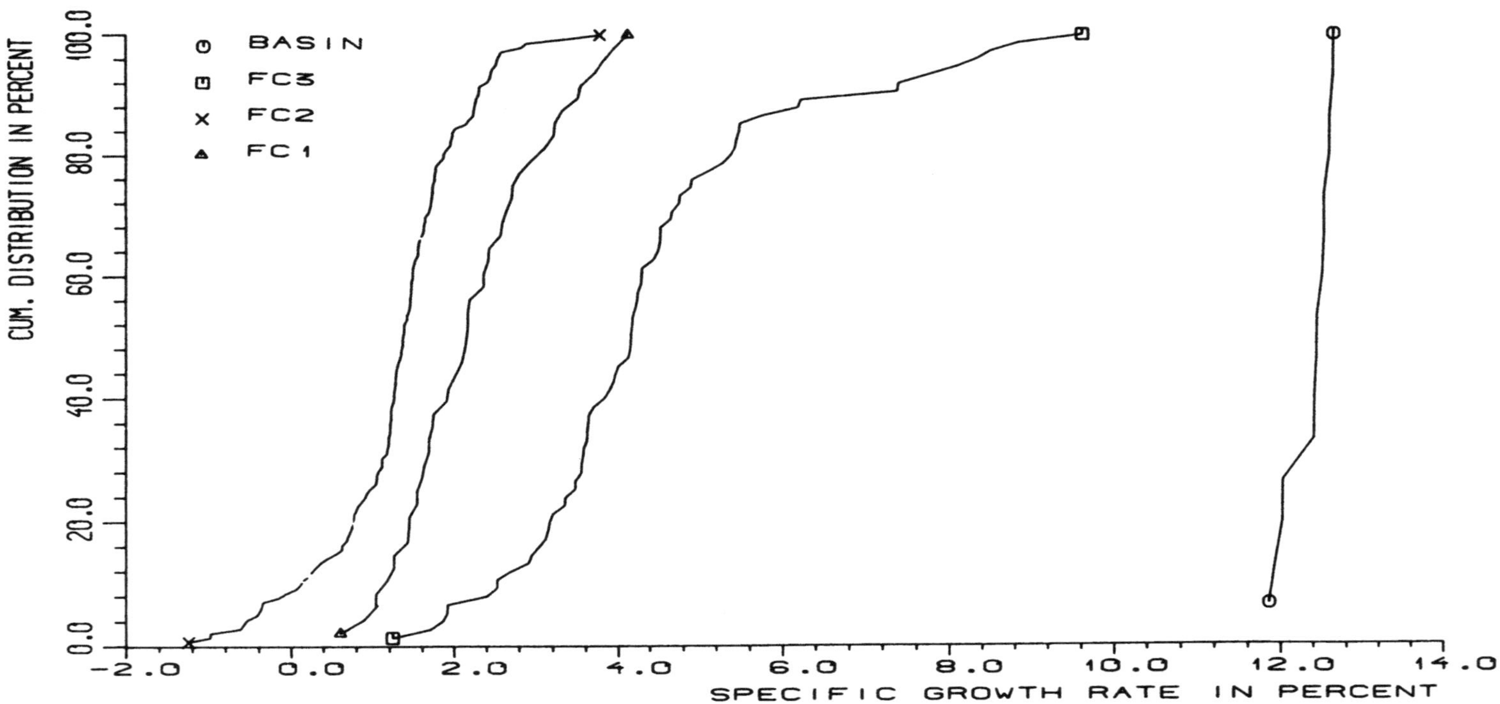

Figure 3. Specific growth rate-frequency distributions for Norwegian herring larvae in 1978, on day 39 in the basin and in the parallel plastic bags (final termination of the bag experiment). FC3 represents the highest food level, FC1 is the lowest level (FC3: n=50; FC2: n=94; FC2: n=32). From Øiestad (1983).

should give a better understanding of the physiology of early life stages of fish. The frequency distribution of SGR in a population or cohort is also of great interest, particularly as it changes as the larvae grow older. Those with the lowest SGR should at any time be those with the highest risk of dying. This is even more likely if there exists a lower limit of SGR for survival beyond metamorphosis. Such a limit would indicate that even larvae which look healthy, and are feeding and growing, might die of starvation. Criteria to detect those larvae could be evolved in meso- and, preferably, in macrocosm studies. These criteria have been dealt with for cod larvae (Øiestad, 1982, 1984), and a minimum SGR for survival beyond metamorphosis has been suggested (Øiestad et al., 1985a). Exact age determination and dry weight measurement of each larva is fundamental for application of this method. Recent progress in age determination (Andersen and Moksness, 1988; Moksness and Wespestad, 1988) should make it possible to apply this method on field-collected material. The minimum SGR has to be corrected for temperature conditions.

The minimum SGR for survival seems to be lower for herring than for cod, indicating a different survival strategy (see Jones, 1973). Particularly in mesocosm studies in plastic bags, herring larvae have survived beyond metamorphosis at low SGR. Autumn-spawned larvae usually survive the winter in a premetamorphosed stage, with very slow growth through the winter, which is part of the opportunistic strategy of herring larvae.

On the other hand, turbot larvae have far higher minimum SGR demands than even cod. Those dying will usually be lost within a fortnight after hatching. A combined macro- and mesocosm study undertaken on turbot larvae revealed the significance of the growth rate by use of daily otolith increments (Rosenberg and Haugen, 1982). This study also gave information on growth trajectories of individual larvae by back-calculations. The ranking of larvae according to length over time was very stable; the few that crossed over represented only minor shifts in the ranking (Fig. 4). This is a very important observation which has also been made on herring larvae (Geffen, 1982) from both macro- and mesocosm studies. One implication of this has been applied in a combined macro- and mesocosm study on herring. A cohort in a plastic bag was divided into three subpopulations based on surviving larvae from three bag terminations in a time series. All larvae were ranked according to their dry weight; the size of the different subpopulations was determined retrospectively (Øiestad, 1983). This strategy, which might be further elaborated, makes it possible to carry out separate calculations for each subpopulation and to obtain a far more exact picture of the dynamics and demands of each group. Among other things, it is possible to define the conditions that result in a particular survival rate for each subpopulation to a particular stage (Øiestad, 1983).

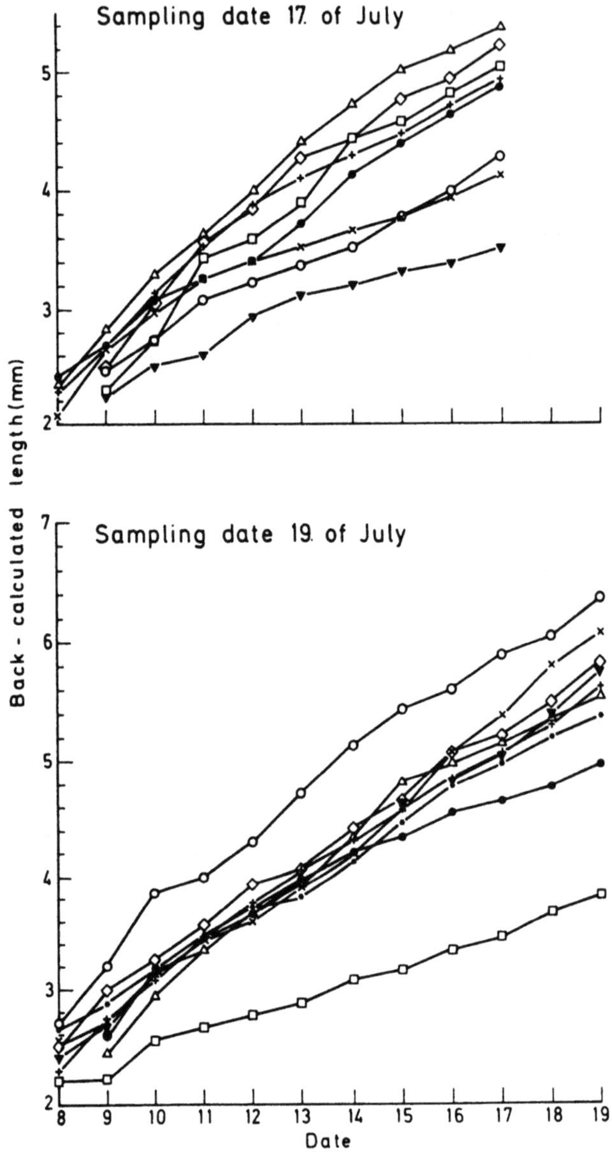

Figure 4. Individual growth trajectories for *Scophthalmus maximus* larvae sampled on the 17th and 19th of July in the basin. The abscissa is the day of the month in July, 1980. From Rosenberg and Haugen (1982).

Great emphasis has been placed on the point of no return (PNR) (May, 1974; Ehrlich et al., 1976; McGurk, 1986). Although PNR is of considerable significance, the growth barrier might be a more useful tool, as the typical situation in the sea might be one where the fish larvae have a food supply and are eating. If the growth barrier works as suggested by Laurence (1977) and Beyer and Laurence (1981), healthy-looking larvae with a too slow growth might be of more significance in the open sea than typically emaciated larvae. Thorough studies of important commercial species including growth barrier and minimum SGR determination should be undertaken, preferably in combined meso- and macrocosm studies. A lot of work has so far been done on cod, herring, capelin, and turbot. But a number of other species should be considered, including those with tropical and subtropical distributions as suggested by Houde (1985).

Predation might be of even greater significance than starvation or too-slow growth, particularly since it operates throughout the whole lifespan. A number of field studies and recent laboratory studies give strong support to predation as a key factor in mortality. However, quantification of this factor meets with methodological obstacles. Fish larvae are difficult to detect in predator guts due to rapid digestion. Furthermore, laboratory studies do not mimic open-sea conditions well due to space; these studies mainly give information on rates of digestion, behavior, satiation, and prey selection. Again, macro- and mesocosm studies should be a powerful tool.

A number of potential invertebrate predators have not been examined. It is difficult to keep some potential predators, like euphausiids, in good shape in enclosures.

However, much has been learned about invertebrate predators during the last ten years. A number of studies with jellyfish indicate that this group causes a significant larval mortality, mainly due to their high density and steady presence in some areas of larval distribution (Gamble and Fuiman, 1987; Fuiman and Gamble, 1988). In some macrocosm studies, jellyfish seem to be responsible for a heavy mortality among cod larvae; as soon as the timing of larval release was changed relative to outbursts of medusae, survival increased dramatically beyond metamorphosis (Kvenseth and Øiestad, 1984; Øiestad, 1985).

With repeated release of cod larvae in macrocosm studies, it is possible to identify different cohorts; and when the first cohort reaches metamorphosis, later released cohorts decline rapidly in number as they probably are exposed to heavy predation (Fig. 1E). This might be the only explanation for this rapid decline of otherwise healthy and fast-growing cod larvae. In most situations, the declines take place at favorable feeding conditions that, according to a large amount of information, should give high survival.

The same phenomenon has been observed repeatedly with delayed release of other fish larvae. With a minor delay, survival has been good for plaice and the plaice-flounder hybrid; longer delay results in extinction of the flatfish larvae when cod larvae metamorphose, although feeding conditions have improved for the flatfish larvae (Øiestad, 1985). With herring and capelin larvae exposed to metamorphosing cod larvae, a sharp decline takes place during cod metamorphosis, although both herring and capelin larvae have had a sufficient food supply and experienced a high SGR (Øiestad, 1983; 1985). Without cod predation, larvae of these two species have high survival rates, even at far lower food concentrations than those observed during their decline when exposed to potential cod predation.

Similar studies have revealed that herring larvae seemingly are able to prey on capelin larvae as soon as the herring start schooling, at a size of about 25 mm. They are even able to hunt for capelin as big as 17 mm. Both this cohort of large and fast-growing capelin as well as a cohort with younger, but also fast-growing larvae were exterminated within a fortnight after herring began schooling. This observation might have considerable ecological significance for the balance of these two species in the Barents Sea, as suggested by Moksness and Øiestad (1987). The overlap of schooling herring with smaller and younger capelin larvae in most of their distribution area along the northernmost coast of Norway in 1983 and later might be one of the main reasons for the collapse of capelin. This study illustrates how macrocosm studies can deal directly with large ecological events, events with tremendous economic significance.

Most of the studies described have had control groups in mesocosms. All these mesocosm studies have strongly supported the assumptions made for the events in the macrocosm studies (Øiestad, 1985).

Interest in postmetamorphosed fish as predators on fish larvae is increasing, and a number of studies have been undertaken in mesocosms. Most of these have had to run for a short time and with low densities of predators. These studies have demonstrated, among other things, the high potential of juvenile clupeids and sand eels for preying on fish larvae even of their own species (Fuiman and Gamble, 1988; Wespestad and Moksness, 1988).

Integrated studies including macrocosms should make these studies far more realistic and give better estimates of predation capacity and rates in an open-sea situation. This might be even more important when dealing with schooling fish species. Even 0-group cod might be considered to be schooling at a prebenthic stage.

Increasing attention among fishery biologists is being directed to inter- and intraspecific interactions like the one illustrated between herring and capelin in the Barents Sea. This interest is strengthened by the reality that fish mainly feed on fish. Nellen (1986) has analyzed this aspect thoroughly, and considers small fish to fill a gap between the huge population of the relatively small zooplankton and larger fish. Small-sized fish can transfer the energy embodied in zooplankton to large fish more efficiently than by other strategies. The enormous fecundity of bony fish compared to other vertebrates could be part of this strategy. Although fish larvae have a very high survival capacity, as has been demonstrated in a number of laboratory and large-scale studies, the ultimate fate for most of them is to be ingested as prerecruits by larger fish. From first feeding until ingestion by a predator, they have harvested the trophic level which would otherwise not be easily available to the fish predator directly. From this point of view, it might be even more important to extend a number of larval studies beyond metamorphosis, both to expose the subjects to predation from larger fish and to let them graze on fish larvae of species normally co-occurring in the open sea. This strategy also would require integrated studies conducted in the laboratory and in meso- and macrocosms.

CONCLUDING REMARKS

Larval studies carried out world-wide since the colloquium at La Jolla in 1975 (Hunter, 1976) have provided the fundamentals for more ambitious studies. A number of the recommendations from that meeting have become subjects of extensive studies and experiments. A few main results have been highlighted here. The need for high food concentrations for larval growth and survival has been completely reconsidered, as laboratory studies on fish larvae obviously do not mimic well a number of important features of the open sea. Food densities of more than 500 organisms per liter in rearing tanks were needed to obtain reasonable growth and survival of a number of marine fish larvae. The same results were obtained in meso- and macrocosms at food densities of 2-10 organisms per liter in the maximum layer of zooplankton. A number of improvements in indoor tank studies lowered the food density required for survival and growth (Houde, 1978; Moffatt, 1981), but there is still a large discrepancy. This might originate partly from microturbulence generated by wind forces in outdoor systems. This turbulence considerably increases the contact rate between predator and prey (Rothschild and Osborn, 1988).

The importance of predation as a regulatory force has been focused on much more, reducing the significance of starvation. As the main predators might be organisms which either hunt randomly, like jellyfish, or have a far higher speed than fish larvae, the theory of selective mortality seems to have less support. Predation on many fish species is most likely undertaken by schooling organisms which leave behind them water masses without any ingestible organisms.

Many fishery biologists prefer to consider the early life stages of fishes as a "black box". They mentally begin with the 0-group stage, as there seems to be a far better relation between number of 0-group and recruitment strength than with number of eggs or larvae. However, the shifting strength of different fish species with overlapping distributions at one or some life-stages might be a result of interaction. The main events might have occurred premetamorphosis, as indicated for the herring-capelin relation in which the capelin stock was reduced to a pale shadow of its earlier strength. Similar, although not so dramatic effects probably are detectable with a number of fish stocks. A deeper understanding should make stock management more predictive.

A large number of studies have contributed to a better understanding of the potential of different types of larval studies. Enough basic information is available for a number of commercial species, such as cod, herring, and capelin which co-exist in eastern and western North Atlantic waters, that it is time for a larger and integrated study incorporating all the methods applied so far.

Instead of looking for the "ideal" enclosure, it might at the present stage be of more interest to combine different types of enclosures and run them synoptically within a macrocosm. That might direct attention to the potential of each system and highlight how to carry out field verification of the observations. As an illustration, the herring-capelin study might initiate continuous monitoring of an important spawning ground for capelin, observing in detail the effect of the arrival of schooling herring fry to this area during hatching of capelin eggs.

There is an obvious need for a number of experts connected with these studies. A specialist within a restricted field might detect phenomena that revolutionize the understanding of dynamics. The enclosure method itself represents in many respects a revolution; detection of the potential effect of microturbulence on predator-prey relations and the importance of highly unsaturated fatty acid (HUFA) (Fraser et al., 1987) for larval growth illustrate the significance of parallel research. It would also be desirable to include traditional stock managers in this work, as field verification will be more important as the level of knowledge grows. Mathematical modelling also is vital. There is a need for long-term dedication to this aspect, as most modelling so far has only been an initiation of something larger (Ellertsen et al., 1981).

LITERATURE CITED

Andersen, T. and E. Moksness. 1988. Estimation of age in days and daily growth rate in larvae and juvenile marine fishes based upon reading daily increments in their otoliths. *ICES 1988 ELHS/*Poster No. 56.

Beyer, J. E. and G. C. Laurence. 1981. Aspects of stochasticity in modelling growth and survival of clupeoid fish larvae. *Rapp. P.-V. Réun. Cons. Perm. Int. Explor. Mer* **178**: 17-23.

Ehrlich, K. F., J. H. S. Blaxter, and R. Pemberton. 1976. Morphological and histological changes during the growth and starvation of herring and plaice larvae. *Mar. Biol.* **35**: 105-118.

Ellertsen, B., E. Moksness, P. Solemdal, S. Tilseth, T. Westgård, and V. Øiestad. 1981. Growth and survival of cod larvae in an enclosure. Experiments and a mathematical model. *Rapp. P.-V. Réun. Cons. Perm. Int. Explor. Mer* **178**: 45-57.

English, K. K. 1983. Predator-prey relationships for juvenile chinook salmon, *Oncorhynchus tshawytscha*, feeding on zooplankton in "in situ" enclosures. *Can. J. Fish. Aquat. Sci.* **40**: 287-297.

Frank, K. T. and W. C. Leggett. 1982. Coastal water mass replacement; its effect on zooplankton dynamics and the predator-prey complex associated with larval capelin (*Mallotus villosus*). *Can. J. Fish. Aquat. Sci.* **39**: 991-1003.

Fraser, A. J., J. R. Sargent, J. C. Gamble, and P. MacLachlan. 1987. Lipid class and fatty acid composition as an indicator of the nutritional condition of larval Atlantic herring. *Am. Fish. Soc. Symp.* **2**: 129-143.

Fuiman, L. E. and J. C. Gamble. 1988. Predation by Atlantic herring, sprat and sand eels on herring larvae in large enclosures. *Mar. Ecol. Prog. Ser.* **44**: 1-6.

Gamble, J. C. 1984. Simultaneous rearing of herring and cod larvae in the Loch Ewe enclosures. *ICES CM 1984/L:* Poster 34 (mimeo.).

Gamble, J. C. and E. D. Houde. 1984. Growth, mortality and feeding of cod (*Gadus morhua* L.) larvae in enclosed water columns and in laboratory tanks. Pp. 123-143. **In**: E. Dahl, D. S. Danielssen, E. Moksness, and P. Solemdal [eds.], *The Propagation of Cod (Gadus morhua L.).* Flodevigen Rapp., 1.

Gamble, J. C., P. MacLachlan, and D. D. Seaton. 1985. Comparative growth and development of autumn and spring spawned Atlantic herring larvae reared in large enclosed ecosystems. *Mar. Ecol. Prog. Ser.* **26**: 19-33.

Gamble, J. C. and L. E. Fuiman. 1987. Evaluation of *in situ* enclosures during a study of the importance of starvation to the vulnerability of herring larvae to a piscine predator. *J. Exp. Mar. Biol. Ecol.* **113**: 91-103.

Geffen, A. J. 1982. Otolith ring deposition in relation to growth rate in herring (*Clupea harengus*) and turbot (*Scophthalmus maximus*) larvae. *Mar. Biol.* **71**: 317-326.

Houde, E. D. 1978. Critical food concentrations for larvae of three species of subtropical marine fishes. *Bull. Mar. Sci.* **28**: 395-411.

Houde, E. D. 1985. Mesocosms and recruitment mechanisms. Council Meeting, *Int. Coun. Explor. Sea 1985, Mini-Symp.*. No. **4**: 1-13 (mimeo.).

Houde, E. D. and S. A. Berkeley. 1982. Food and growth of juvenile herring *Clupea harengus pallasi*, in CEPEX enclosures. Pp. 239-250. **In**: G. D. Grice and M. R. Reeve [eds.], *Marine Mesocosms. Biological and Chemical Research in Experimental Ecosystems.* New York: Springer-Verlag.

Hunter, J. R. 1976. Report of a colloquium on larval fish mortality studies and their relation to fishery research, January 1975. *U.S. NOAA, Tech. Rep., NMFS Circular.* No. **395**: 1-5.

Jones, R. 1973. Density dependent regulation of the numbers of cod and haddock. *Rapp. P.-V. Réun. Cons. Perm. Int. Explor. Mer* **164**: 156-173.

Koeller, P. and T. R. Parsons. 1977. The growth of young salmonids (*Oncorhynchus keta*): controlled ecosystem pollution experiment. *Bull. Mar. Sci.* **27**: 114-118.

Kvenseth, P. G. and V. Øiestad. 1984. Large-scale rearing of cod fry on the natural food production in an enclosed pond. Pp. 645-655. **In**: E. Dahl, D. S. Danielssen, E. Moksness, and P. Solemdal [eds.], *The Propagation of Cod (Gadus morhua L.).* Flodevigen Rapp., **1**.

de Lafontaine, Y. and W. C. Leggett. 1987. Evaluation of *in situ* enclosures for larval fish studies. *Can. J. Fish. Aquat. Sci.* **44**: 54-65.

Laurence, G. C. 1977. A bioenergetic model for the analysis of feeding and survival potential of winter flounder, *Pseudopleuronectes americanus*, larvae during the period from hatching to metamorphosis. *U.S., Fish. Wildl. Serv., Fish. Bull.* **75**: 529-546.

Laurence, G. C., T. A. Halavik, B. R. Burns, and A. S. Smigielski. 1979. An environmental chamber for monitoring *in-situ* growth and survival of larval fishes. *Trans. Am. Fish. Soc.* **108**: 197-203.

Leggett, W. C. 1986. The dependence of fish larvae survival on food and predator densities. Pp. 117-138. **In**: S. Skreslet [ed.], *The Role of Freshwater Outflow in Coastal Marine Ecosystems*. New York: Springer-Verlag.

May, R. C. 1974. Larval mortality in marine fishes and the critical period concept. Pp. 3-19. **In**: J. H. S. Blaxter [ed.], *The Early Life History of Fish.* Berlin: Springer-Verlag.

McGurk, M. D. 1986. Natural mortality of marine pelagic fish eggs and larvae: role of spatial patchiness. *Mar. Ecol. Prog. Ser.* **34**: 227-242.

Moffatt, N. M. 1981. Survival and growth of northern anchovy larvae on low zooplankton densities as affected by the presence of a *Chlorella* bloom. *Rapp. P.-V. Réun. Cons. Perm. Int. Explor. Mer* **178**: 475-480.

Moksness, E. 1982. Food uptake, growth and survival of capelin larvae (*Mallotus villosus* Muller) in an outdoor constructed basin. *Fish. Dir. Skr. Ser. HavUnders.* **17**: 267-285.

Moksness, E. and V. Øiestad. 1987. Interaction of Norwegian spring-spawning herring larvae (*Clupea harengus*) and Barents Sea capelin larvae (*Mallotus villosus*) in a mesocosm study. *J. Cons. Int. Explor. Mer* **44**: 32-42.

Moksness, E. and V. Wespestad. 1988. Ageing and back-calculating growth rate of Pacific herring (*Clupea harengus pallasi*) larvae by reading daily increments in their otoliths. *ICES 1988 ELHS/* Poster No. 30.

Nellen, W. 1986. A hypothesis on the fecundity of bony fish. *Ber. Dt. Wiss. Kommn. Meeresforsch.* **31**: 75-89.

Øiestad, V. 1982. Application of enclosures to studies on the early life history of fishes. Pp. 49-62. **In**: G. D. Grice and M. R. Reeve [eds.], *Marine Mesocosms. Biological and Chemical Research in Experimental Ecosystems.* New York: Springer-Verlag.

Øiestad, V. 1983. Growth and survival of herring larvae and fry (*Clupea harengus* L.) exposed to different feeding regimes in experimental ecosystems: outdoor basin and plastic bags. Unpubl. thesis (Dr. Phil.): University of Bergen. 299 pp.

Øiestad, V. 1984. Criteria for condition evolved from enclosure experiments with cod larva populations. Pp. 213-229. **In**: E. Dahl, D. S. Danielssen, E. Moksness, and P. Solemdal [eds.], *The Propagation of Cod (Gadus morhua L.).* Flodevigen Rapp., 1.

Øiestad, V. 1985. Predation on fish larvae as a regulatory force, illustrated in mesocosm studies with large groups of larvae. *NAFO Sci. Counc. Stud.* **8**: 25-32.

Øiestad, V., B. Ellertsen, P. Solemdal, and S. Tilseth. 1976. Rearing of different species of marine fish fry in a constructed basin. Pp. 303-329. **In**: G. Persoone and E. Jaspers [eds.], *Proc. 10th Eur. Mar. Biol. Symp.*, Vol. 1. Wettern, Belgium: Universa Press.

Øiestad, V. and V. Moksness. 1981. Study of growth and survival of herring larvae (*Clupea harengus* L.) using plastic bag and concrete basin enclosures. *Rapp. P.-V. Réun. Cons. Perm. Int. Explor. Mer* **178**: 144-152.

Øiestad, V., A. Folkvold, and P. G. Kvenseth. 1985a. Growth-pattern of Atlantic cod larvae (*Gadus morhua* L.) from first feeding to metamorphosis studied in a mesocosm. Council Meeting, *Int. Coun. Explor. Sea, 1985, Mini-Symp.* No. **9**: 1-10 (mimeo.).

Øiestad, V., P. G. Kvenseth, and A. Folkvold. 1985b. Mass production of Atlantic cod juveniles (*Gadus morhua* L.) in a Norwegian saltwater pond. *Trans. Am. Fish. Soc.* **114**: 590-595.

Rognerud, C. 1887. Hatching cod in Norway. *Bull. U.S. Fish Commn.* **8**: 113-119.

Rollefsen, G. 1946. Kunstig oppdrett av flyndreyngel. Pp. 91-113. **In**: C. L. Godske [ed.], *Forskning og Framsteg.* Bergen: J. W. Eides Forlag.

Rosenberg, A. A. and A. S. Haugen. 1982. Individual growth and size-selective mortality of larval turbot (*Scophthalmus maximus*) reared in enclosures. *Mar. Biol.* **72**: 73-77.

Rothschild, B. J. and T. R. Osborn. 1988. Small-scale turbulence and plankton contact rates. *J. Plankton Res.* **10**: 465-474.

Schnack, D. 1981. Studies on the mortality of Pacific herring larvae during their early development using artificial in-situ containments. *Rapp. P.-V. Réun. Cons. Perm. Int. Explor. Mer* **178**: 135-142.

Wespestad, V. G. and E. Moksness. 1990. Observations on the early life history of Pacific herring (*Clupea pallasi*) from Bristol Bay, Alaska, in a marine mesocosm. *Fish. Bull.* **88**.

9. APPLICATION OF MESOCOSMS FOR SOLVING PROBLEMS IN POLLUTION RESEARCH

Michael E. Q. Pilson

Abstract

Much of the research with marine mesocosms appears to have been driven by a need to investigate the fates and effects of pollutants, with the consequence that basic underlying information on the principles that govern the functioning of these experimental systems and of the natural systems of which they are living models has been given inadequate attention. Nevertheless, research with mesocosms of several types has demonstrated that this approach has great potential to help unravel the processes that affect both the fates of pollutants and their effects on complex ecosystems. Some kinds of information would have been very difficult or impossible to obtain without experimental research in these complex living models. In addition, this research has provided a great deal of basic ecological information about marine ecosystems.

INTRODUCTION

When undertaking to prepare this paper I felt at first uneasy about the subject matter. Examination of the rather extensive literature (e.g. the dedicated issue in 1977 of *Bull Mar. Sci.* 27: 1-175; the symposium volumes edited by Giesey, 1980, and Grice and Reeve, 1982; and a review of the use of aquatic microcosms in ecotoxicological research by Gearing, 1989, with some 400 references) suggests that much of the work with marine mesocosms has been carried out with funds oriented toward the study of pollution. Many pollutants, several types of ecosystems, and many approaches have been used. This literature, however, is not characterized by generality and rigor of approach. Many investigators have felt compelled to address short-term responses to perturbations without a full definition of the systems with which they are working. The experimental use of enclosed marine ecosystems to study pollutants needs to be built on a stronger foundation.

The study of pollution in the marine environment ought to be a branch of marine ecology. In an ideal world, the behavior and fates of pollutants would be considered in the light of some understanding of the natural fluxes and pathways of the substances of concern (or, in the case of entirely man-made substances, the pathways of like substances), and their effects would be considered only after we understand or are at least aware of the responses of the ecosystem to natural variations in the parameters which are important in controlling its expression. The use of experimental ecosystems in the study of marine pollution would be most effectively advanced by further progress in understanding the properties of the systems used and of the natural ecosystems to which they relate.

Whereas there were a number of important and seminal earlier investigations, most of the work with marine mesocosms has taken place during the last 20 years. During this time, there is no entirely satisfying example where the properties of the mesocosm have been described thoroughly enough that the potential longevity of the system, its differences from the "natural" system, and the reasons for those differences, have been thoroughly worked out.

We may suppose that, depending on its nature, the behavior of each enclosed ecosystem or mesocosm will change in some fashion in response to such variables as the size and/or shape of the enclosure, the total light energy input and its spatial and temporal characteristics, the temperature and its seasonality, the time of set-up and the duration of the enclosure, the nature and amount of allochthonous organic matter input, the energy and structure of the turbulence field in the enclosure, the type of sediment present (if any), and the vast range of nutrient and other trace chemical constituents varying individually, synergistically and temporally, Whereas some information exists on most of the items in the list above, obtained from one system or another, often in some illuminating detail, I am not aware of any rigorously controlled experiment from which we may conclude the sensitivity of any marine mesocosm to the variables mentioned. Nor have I found a completely satisfactory comparison of the behavior of any mesocosm with that of the ecosystem from which it is derived. No mesocosm could ever behave identically to the natural ecosystem, if only because the natural system is usually patchy, spatially and temporally variable, possibly chaotic, and responds to far-field influences. A thorough documentation of both similarities and differences would help immensely to provide insight into the sensitivity of both the natural system and the mesocosm to the controlling variables, and would greatly strengthen conclusions that may be obtained from experiments with pollutants.

We do not live in an ideal world, however, and we will certainly continue to experiment with pollutants in mesocosms, even without fully exploring their unpolluted behavior. By doing so, we will continue to learn a great deal about the natural behavior of these systems, as well as about the fates and effects of pollutants.

DEFINITION OF A MESOCOSM

A distinction must be made between an aquarium and a mesocosm. In an aquarium, the organisms of interest are generally supplied their food, are maintained at concentrations usually much greater than could exist naturally, and they may not even reproduce. In a mesocosm, the intention is usually that the organisms are at natural concentrations and live on the food generated within the mesocosm itself. Exceptions may be mesocosms of benthic systems that naturally require allochthonous inputs of food, and arrangements are usually made to supply this. I consider that a mesocosm is a living model of a complex ecosystem. It is an enclosure that has been set up so as to capture some portion of the many interacting species and complex processes of a natural ecosystem, and the system should be self-maintaining.

If the source of metabolic energy for the system is photosynthesis, as is usually the case for a planktonic system, then the mesocosm should be supported entirely by its own photosynthesis. If the system is benthic exclusively, then arrangements would usually be made to supply the organic input so as to replace what would be the input from the planktonic region above, in addition to any photosynthesis on the bottom. If both benthos and plankton are present, then presumably the benthos should live on what the plankton provide.

The question of what is meant by "self-maintaining" is not really answered. If the system begins to change greatly as soon as it is enclosed, then this should point immediately to an investigation of those important parameters that we do not yet understand. On the other hand, we can be reasonably sure that no enclosure can ever maintain a complex population just as in some natural system. Always there must be some balance struck, but we yet lack an agreement or even a full discussion on how to strike a balance, or even on how to think about these matters.

GOALS OF POLLUTION RESEARCH

In the study of any pollutant, it is well to keep in mind the distinction between the fate of a substance and the effects of that substance. The first thing is to learn how much is introduced, where the substance goes, in what ways it may be transformed or trapped, and what the ultimate exposure concentrations are likely to be in the ecosystem of concern. Whereas advective transport can only be known from field measurements and modelling, the processes of transformation and trapping are appropriately studied in model ecosystems. Substances may be biologically, chemically, or photochemically modified in the marine environment, or they may be volatilized, complexed, adsorbed to particles, transported on particles, and very often sequestered in some fashion in the sediments. All these processes are appropriately studied by some combination of experimentation in the laboratory and in living models of the ecosystem (mesocosms). Only with some realistic understanding of the fate of a substance is it appropriate to examine the effects.

Pollutants are considered to be pollutants if it is thought that they may cause some damage to individuals, to species, or to the whole ecosystem. There is a long history of investigation on the toxic properties of substances when single species are exposed to them. As with the use of rats and mice in medical practice, these studies will continue to provide the foundation of our knowledge of the physiological effects of all substances. In considering the effect that pollutant substances may have on the whole ecosystem, however, it is clear that we can never study individually all the species of organisms present. Furthermore, test organisms tend to be the most hardy and adaptable to laboratory culture. It is thought that there may often be more sensitive species in the ecosystem, and these species must often interact in ways we are yet unaware of. The way to detect this is to do experimental studies on the whole system.

A practical problem arises, however, when considering doing the experiments on enclosed ecosystems in mesocosms to discover the effects of introduced substances. We commonly do not know what to measure. If, as is sometimes advocated to save money and time, we measure general properties of the system (such as respiration, photosynthesis, or pH), it may be imagined that sensitive species die, and others take their place, and we will never know it. On the other hand, if we attempt to keep track of the numbers of each species present, then every test is very expensive and time-consuming, and the experiment may have to last long enough to include the complete life cycles of all important organisms. At the present time, there is no general resolution of this dilemma, and there may never be. The nature of the evidence looked for in each case must lead to the measurement made.

In the following sections, I consider briefly a few selected examples which illustrate studies of both fates and effects of pollutant substances.

EXAMPLES OF STUDIES OF FATES

Organic Compounds

At the Marine Ecosystems Research Laboratory (MERL), University of Rhode Island, a set of 14 tanks is run as a living model of a coupled planktonic and benthic ecosystem typical of Narragansett Bay, and to some extent of the shallow coastal waters in much of the temperate North Atlantic. These mesocosm tanks usually contain 13 m^3 of water and a ton of sediment, and they naturally maintain chemical concentrations and process rates, and biological components, quite reminiscent of those in Narragansett Bay (Vargo et al., 1982). Since the hundreds of biological and biogeochemical processes that all together result in poising the concentration of phosphate (Fig. 1) seem to result in near-natural concentrations of this element (and of other nutrients), it was concluded (Pilson, 1985) that the major processes that control these substances had been captured in these tanks. It seemed a satisfactory working hypothesis that most other biogeochemical processes might be equally well represented, and at least some are. This provides some confidence in the results from investigations of the fates of pollutants in these systems.

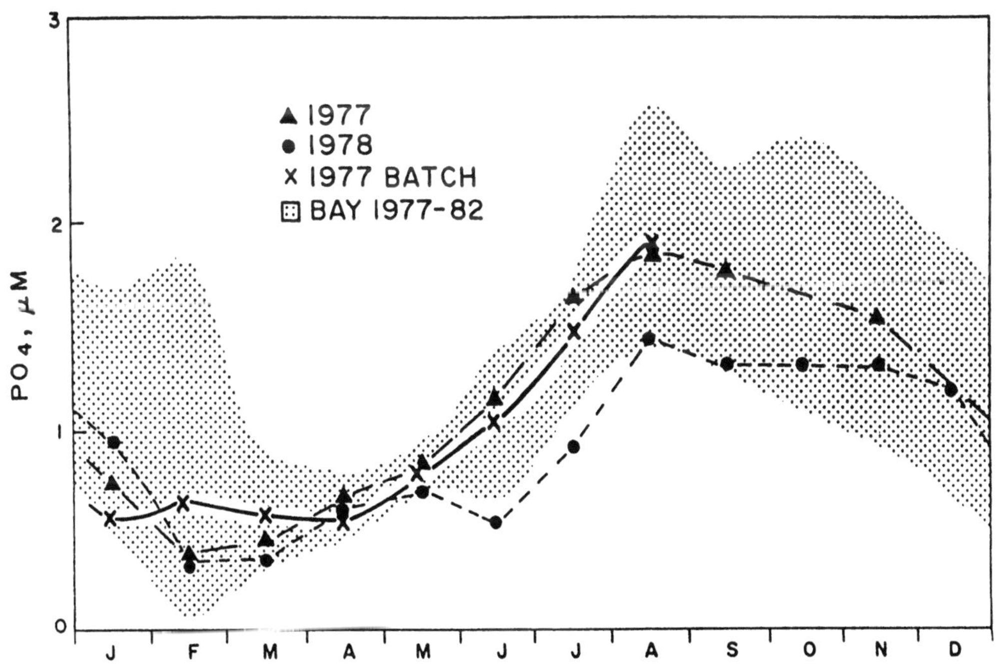

Figure 1. Concentration of dissolved inorganic phosphate in lower Narragansett Bay and MERL mesocosms. The stippled area represents the field in which were found 95% of the monthly means of weekly measurements of the concentration of phosphate at the GSO dock. Circles and triangles represent similar means for three control mesocosms during two consecutive years of operation in a flow-through mode (water replacement rate, about 3.7% per day) and the X's represent the means of three tanks operated in batch mode, with no water replacement. From Pilson (1985).

Several experiments on the fates of various unlabelled and labelled organic compounds have been carried out in the MERL tanks. Wakeham and co-workers (Wakeham et al., 1982, 1983, 1985, 1986a, 1986b) investigated the fates of about twenty volatile organic compounds at beginning concentrations of only about 0.3 to 3 $\mu g\ l^{-1}$, overlapping the range of concentrations observed for these compounds in parts of Narragansett Bay. Comparing the rates of disappearance of degradable compounds with those of largely inert (heavily chlorinated) compounds, the rates of loss by degradation could be distinguished from rates of loss by volatilization. Rates of degradation were very slow in the winter, and much faster (half-lives sometimes less than 1 day) in summer. An interesting observation was that loss rates sometimes changed greatly over a very short (1-2 days) interval of

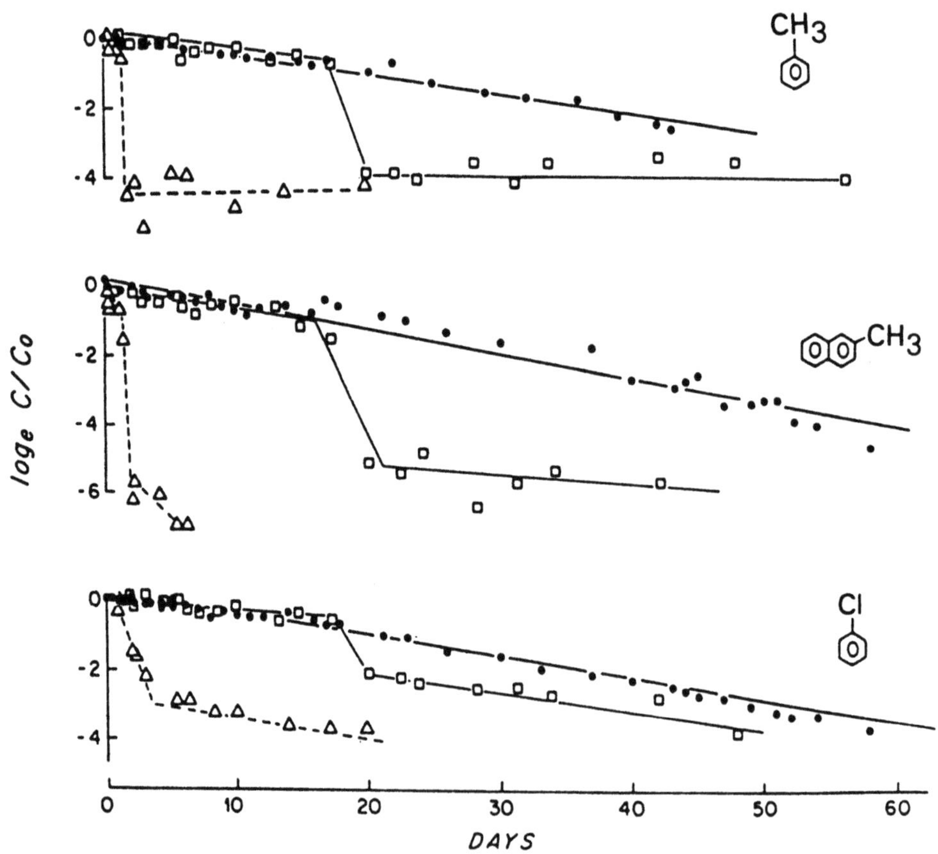

Figure 2. Concentration changes with time for toluene, 2-methylnaphthalene, and chlorobenzene in MERL tanks. Closed circles, winter experiments; squares, spring; triangles, summer. Initial concentrations ranged from 3.1 to 3.6 $\mu g\ l^{-1}$ for toluene, 0.5 to 0.9 $\mu g\ l^{-1}$ for 2-methylnaphthalene, and 0.6 to 2.8 $\mu g\ l^{-1}$ for chlorobenzene. From Wakeham et al. (1983). Reprinted with permission from *Environmental Science & Technology*, Vol. 17. Copyright (1983) American Chemical Society.

time (e.g. Fig. 2). In one case, the sudden increase coincided with a poorly documented bloom of phytoplankton. This increased rate was ascribed to the development of some bacterium capable of metabolizing the compounds involved (toluene, chloronaphthalene, and chlorobenzene). It is not known whether some bacterium "learned" to metabolize these compounds, or just came along by chance and was able to multiply because of the phytoplankton bloom. We need more studies of this phenomenon. In some experiments, toluene was labelled with radiocarbon and the label followed in some detail, quantifying at each sampling time during a two-month experiment the amount of label in toluene, released as CO_2, and bound in particulate matter.

Thus these experiments demonstrated that, in mesocosm-type systems, it is possible to learn a great deal about the detailed behavior (rates of volatilization, rates of biological degradation and transformation) and the effects of ecosystem functions and ecosystem variability on these rates. Under some circumstances, it is possible to study many compounds simultaneously, and at concentrations unlikely to have much effect on the ecosystem itself.

A series of experiments with ^{14}C-labelled benzanthracene and dimethyl benzanthracene in the MERL tanks was carried out by Hinga and co-workers (Hinga et al., 1980, 1986, Hinga and Pilson, 1987). In these experiments, the amount of benzanthracene added was less than the amount (approximately 5 mg ton^{-1}) already present in the sediments as collected from Narragansett Bay. The same experiments continued longer than 200 days. In the case of benzanthracene (the more stable compound), the amount of parent compound in the water column decreased to negligible amounts in about 50 days, and several water-soluble but extractable fractions appeared, lasted much longer, and for much of the time were present in greater concentrations than benzanthracene (Fig. 3). In addition (not shown), one or more water-soluble compounds, not extractable by chloroform, amounted to as much as 30% of the total label added even after 100 days. Labelled benzanthracene that was sequestered into the sediments appeared to be protected, and little or no degradation occurred between day 120 and day 202.

From these observations it is evident that merely examining the disappearance of some compound will often tell us little about what may happen to it. The long term persistence of benzanthracene in the sediments suggests that once sequestered there, it is inert. The number and apparent stability of compounds initially formed from benzanthracene and appearing in the water column suggest that it is important to examine the degradation products of organic chemical pollutants. In the case of benzanthracene, we do not yet know what these degradation products are. We do know that they are not the same compounds formed by metabolism of benzanthracene in mammalian liver. We do not yet know how to search for these compounds in nature; they were discovered only because the experiment was carried out with benzanthracene labelled with a radiotracer. Nor do we know whether they are more, or less, toxic than benzanthracene itself. The information learned so far, however, could only have been obtained in its present state of reliability by experimentation in such living model ecosystems.

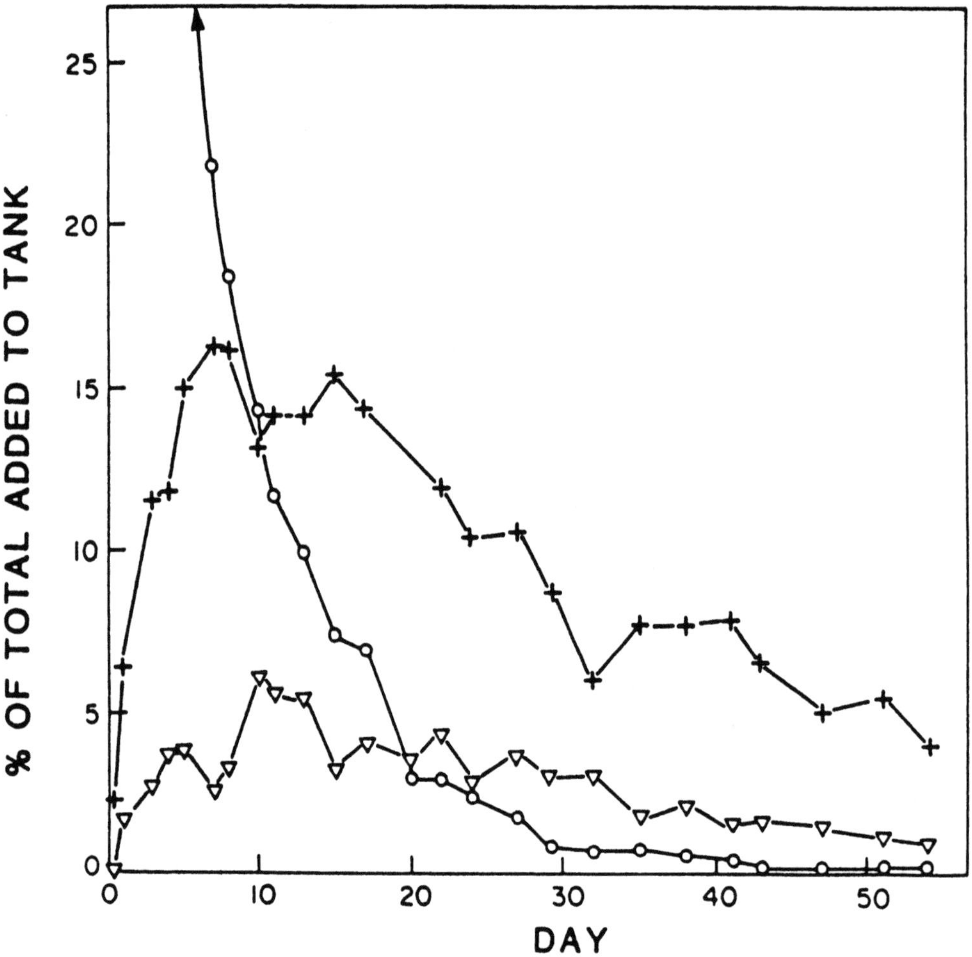

Figure 3. Results of an experiment with 3.5 mg of ^{14}C-labelled benzanthracene added to a MERL tank, giving an initial concentration of 270 ng l^{-1}. Open circles show the concentration of benzanthracene itself, triangles show the concentration of a single compound "band 1" that is extractable with chloroform and runs separately from benzanthracene on a thin-layer chromatogram. Crosses show the concentration of chloroform-extractable material that remains at the origin. Not shown are the concentrations of one or more long-lived products that were not extractable with chloroform and by day 100 amounted to more than 30% of the total activity added to the tank. From Hinga and Pilson (1987). Reprinted with permission from *Environmental Science & Technology*, Vol. 21. Copyright (1987) American Chemical Society.

Fates of Metals

There have been many studies of the fates of many metals and other trace elements introduced into a variety of marine mesocosms (e.g. Santschi, 1988; Gearing, 1989). The unique advantages of mesocosm experimentation with metals and other elemental pollutants are similar to those for organic compounds, except that while often subject to transformation, they are not permanently destroyed.

Here I call attention to the use of temporary enclosures to examine the fates of introduced metals. The Bremerhaven caissons (Schulz-Baldes et al., 1983; Farke et al., 1984) provide an effective way to examine at least part of the biogeochemical behavior of added trace elements. These constructions, enclosing 13 m^2 of sediment, were placed on the shallow sand flats near Norderney in the German Bight, and were allowed to fill and empty with the tide. For several weeks during each tidal filling, trace metals were introduced at measured rates, and the concentrations measured in water, sediment, and benthic organisms. Later, the loss rates of the accumulated metals were measured. This approach of using temporary enclosures has evident advantages. The organisms and sediment are initially not much perturbed by the experimental setup. The water and its dissolved and particulate concentrations are renewed fresh on each tide. The organisms presumably feed naturally (but this should be ascertained). Uptakes of trace metals into organisms and sediment are measured in a system operating in a nearly natural fashion.

Disadvantages are that the changes of the benthic ecosystem within the enclosure are perhaps to be expected, so that the exploration of longer-term effects must be carefully considered, and any slow chemical diagenetic processes in the water column taking longer than a few hours cannot be investigated in these experiments. Turbulent transport processes may need to be examined so that rates inside the enclosures can be related to expected rates in the field outside. The disadvantages seem small, and the use of such temporary enclosures, appropriately adapted to each local environment, will provide continuing opportunities for further experimental work.

EXAMPLES OF STUDIES OF EFFECTS

One of the early uses of the big bag planktonic mesocosms (the CEPEX and Loch Ewe experiments) was to examine the effects of added heavy metals. In the Loch Ewe experiment (Davies and Gamble, 1979), the effect of mercury added at 1 $\mu g\ l^{-1}$ was transient and even uncertain, whereas at 10 $\mu g\ l^{-1}$ there was a marked reduction of the zooplankton which also led to further effects on the ecosystem. These results were similar to those of experiments in the larger CEPEX enclosures, and together were a demonstration of complex results propagating from one affected component of the ecosystem to another.

The effects of oil pollution have been studied in considerable detail, both in the field around natural seeps and after accidental spills, and in mesocosms. As Howarth (1989) said, "The mesocosm approach has proved powerful," and results from such experimental work have helped elucidate the effects expected in the field, as well as providing insight into the probable complexity of these effects.

A demonstration of complex results observed in mesocosm experiments, but that could not have been predicted from single-species bioassays, came from a series of additions of water-accommodated fuel oil at concentrations of 90 and 180 $\mu g\ l^{-1}$ (Grassle et al., 1981; Oviatt et al., 1982). At the higher levels of oil addition the tanks looked green and healthy, and phytoplankton were much more abundant than in the controls. This was evidently due to the greater sensitivity of the zooplankton and of some benthic filter feeders, reducing the grazing pressure on the phytoplankton. Effects on the population structure of the phytoplankton were, however, relatively minor. In the benthos, ciliates and foraminifera were more abundant than in the controls, while amphipods and many other crustacea were eliminated. There, effects were dramatic, and could not have been predicted in detail with confidence from any other source of information. Elmgren and Frithsen (1982) showed that the results of these experiments were in agreement with observations at the site of the Tsesis oil spill in the Baltic.

An experiment which described the effects of an oil mousse introduced for a short time onto experimental muddy tidal flats (Kuiper et al., 1984) was notable for the care taken to set up the mesocosms, the length of the pre-experimental period, and the degree of replication achieved. Significant observations included the behavior of the oil (which appeared to realistically modelled in these enclosures), the rate of burial of oil in the sediment, the unexpectedly long period of time over which the oil kept coming back out from the sediments, and the biological effects which must have been due in part to the prolonged exposure of organisms to significant levels of released oil in the water column. Experiments such as these have provided a firm experimental foundation upon which to base an understanding of both the behavior and the effects of oil in the environment, and they also suggest productive avenues for further research.

DISCUSSION

The difficulties of detecting the effects of individual pollutant substances in nature are compounded by the normal variability of the natural system, as well as by the usual situation that many substances are commonly discharged together and may follow different biogeochemical pathways upon dispersion and dilution. The first concern is not entirely easy to deal with, because mesocosms are often naturally variable, but the possibility of repeated and replicated experimentation means that cause and effect can be reliably established. The second concern is ideally dealt with in mesocosm research by adding substances one at a time, or in known combination.

It may be argued that if mesocosms were not naturally variable, they would be unnatural. The imposition of sufficient control to prevent variability and apparent lack of close replicability would only be removing in some way much of the normal composition of the system. For some experimental purposes this may be desirable, but in general we don't know how to do it. We may often be able to capture replicate samples of sufficiently homogeneous or mixed part of some ecosystem that they will appear to be very similar. Then we can carry out short-term experiments before too much divergence occurs. On the other hand, if we wish to maintain ecosystems for a considerable period of time, so as to observe results through the entire life cycles of organisms bigger than phytoplankton, or seasonally dependent effects, we have to accept natural divergence and variability.

Given the inherent variability of living ecosystems in mesocosms, the question of replicability and the detection of effects becomes important. In cases where the effects are dramatic, there should be little difficulty; but investigation of more subtle effects may require more attention to the natural variability of mesocosm systems, as well as to the natural variability of nature, than has so far been common.

Most of the available examples of investigations into the fates of substances as studied in mesocosms come from temperate environments. It might be expected that results would be different under polar or tropical conditions, but I am aware of no extensive experiments on pollutants in mesocosms set up to simulate either polar or tropical conditions. The experiments of Wakeham (e.g. Fig. 2) showed that rates of degradation were very different in the summer and the winter, so that in the summer most of the loss of some compounds studied was through biodegradation, whereas in the winter most of the loss was through volatilization. One might expect that organic compounds might be degraded much faster under really tropical conditions, but we have no experiments to prove it. Similarly, polar fauna and flora might be adapted to metabolize compounds faster than temperate organisms in the winter but, again, we have no mesocosm experiments that address this.

SUMMARY

During the last two decades, much of the work with mesocosms has been directed toward the study of the fates and effects of pollutants, while perhaps inadequate attention has been paid to the ecological processes at work in these systems and in nature.

From this work, we have learned a great deal about the biogeochemical pathways and fates of many trace elements and of numerous organic compounds in temperate coastal waters. For example, in many cases we now have quantitative information on rates of degradation and on rates at which substances are sequestered into the sediments.

Experiments of the effects of numerous substances in a variety of mesocosms have shown that in many cases the effects could not have been predicted, because of the complexity of interaction between the many species present. Again, most of the work has been in mesocosms of temperate coastal systems, and mostly in the summertime.

Our understanding of both the fates and the effects of pollutant substances would be greatly enriched if studies could be carried out in parallel in mesocosms in polar and tropical regions.

The mesocosm approach has proved very useful, and we have learned a great deal about the effects of pollutants that could not have been learned any other way, or could have been learned only with great difficulty. The experiments have also provided a much more secure knowledge about how to experiment with enclosed ecosystems in mesocosms, and in so doing have given considerable insight into the factors which are important in affecting and controlling the larger natural systems from which the mesocosms are taken.

LITERATURE CITED

Davies, J. M. and J. C. Gamble. 1979. Experiments with large enclosed ecosystems. *Phil. Trans. R. Soc. London* B **286**: 523-544.

Elmgren, R. and J. B. Frithsen. 1982. The use of experimental ecosystems for evaluating the environmental impact of pollutants: A comparison of an oil spill in the Baltic Sea and two long-term, low-level oil addition experiments in mesocosms. Pp. 153-165. **In**: G. D. Grice and M. R. Reeve [eds.], *Marine Mesocosms. Biological and Chemical Research in Experimental Ecosystems.* New York: Springer-Verlag.

Farke, H., M. Schulz-Baldes, K. Ohm, and S. A. Gerlach. 1984. Bremerhaven caisson for intertidal field studies. *Mar. Ecol. Prog. Ser.* **16**: 193-197.

Gearing, J. N. 1989. The role of aquatic microcosms in ecotoxicologic research as illustrated by large marine systems. Pp. 411-472. **In**: S. A. Levin, M. A. Harwell, J. R. Kelly, and K. D. Kimball [eds.], *Ecotoxicology: Problems and Approaches*. New York: Springer-Verlag.

Giesy, J. P., Jr. [ed.]. 1980. *Microcosms in Ecological Research*. DOE Symposium Series 52, CONF-781101. Springfield, VA: National Technical Information Service. 1110 pp.

Grassle, J. F., R. Elmgren, and J. P. Grassle. 1981. Response of benthic communities in MERL experimental ecosystems to low level, chronic additions of #2 fuel oil. *Mar. Environ. Res.* **4**: 279-297.

Grice, G. D. and M. R. Reeve [eds.]. 1982. *Marine Mesocosms. Biological and Chemical Research in Experimental Ecosystems.* New York: Springer-Verlag. 430 pp.

Hinga, K. R., M. E. Q. Pilson, R. F. Lee, J. W. Farrington, K. Tjessem, and A. C. Davis. 1980. Biogeochemistry of benzanthracene in an enclosed marine ecosystem. *Environ. Sci. Technol.* **14**: 1136-1143.

Hinga, K. R., M. E. Q. Pilson, G. Almquist, and R. F. Lee. 1986. The degradation of 7, 12-dimethylbenz(a)anthracene in an enclosed marine ecosystem. *Mar. Environ. Res.* **18**: 79-91.

Hinga, K. R. and M. E. Q. Pilson. 1987. Persistence of benz[a]anthracene degradation products in an enclosed marine ecosystem. *Environ. Sci. Technol.* **21**: 648-653.

Howarth, R. W. 1989. Determining the ecological effects of oil pollution in marine ecosystems. Pp. 69-97. **In**: S. A. Levin, M. A. Harwell, J. R. Kelly, and K. D. Kimball [eds.], *Ecotoxicology: Problems and Approaches*. New York: Springer-Verlag.

Kuiper, J., P. A. W. J. de Wilde, and W. J. Wolff. 1984. Oil pollution experiment (OPEX). I. Fate of an oil mousse and effects on macrofauna in a model ecosystem representing a Wadden Sea tidal mudflat. Pp. 331-359. **In**: G. Persoone, E. Jaspers, and C. Claus [eds.], *Ecological Testing for the Marine Environment*. Vol. 2. State Univ. of Ghent.

Oviatt, C. A., J. Frithsen, J. Gearing, and P. Gearing. 1982. Low chronic additions of No. 2 fuel oil: chemical behavior, biological impact and recovery in a simulated estuarine environment. *Mar. Ecol. Prog. Ser.* **9**: 121-136.

Pilson, M. E. Q. 1985. Annual cycles of nutrients and chlorophyll in Narragansett Bay, Rhode Island. *J. Mar. Res.* **43**: 849-873.

Santschi, P. H. 1988. Factors controlling the biogeochemical cycles of trace elements in fresh and coastal marine waters as revealed by artificial radioisotopes. *Limnol. Oceanogr.* **33**: 848-866.

Schulz-Baldes, M., E. Rehm, and H. Farke. 1983. Field experiments on the fate of lead and chromium in an intertidal benthic mesocosm, the Bremerhaven caisson. *Mar. Biol.* **75**: 307-318.

Vargo, G. A., M. Hutchins, and G. Almquist. 1982. The effect of low, chronic levels of No. 2 fuel oil on natural phytoplankton assemblages in microcosms. 1. Species composition and seasonal succession. *Mar. Environ. Res.* **6**: 245-264.

Wakeham, S. G., A. C. Davis, and J. T. Goodwin. 1982. Biogeochemistry of volatile organic compounds in marine experimental ecosystems and the estuarine environment - initial results. Pp. 137-151. **In**: G. D. Grice and M. R. Reeve [eds.], *Marine Mesocosms. Biological and Chemical Research in Experimental Ecosystems.* New York: Springer-Verlag

Wakeham, S. G., A. C. Davis, and J. L. Karas. 1983. Mesocosm experiments to determine the fate and persistence of volatile organic compounds in coastal seawater. *Environ. Sci. Technol.* **17**: 611-617.

Wakeham, S. G., E. A. Canuel, P. H. Doering, J. E. Hobbie, J. V. K. Helfrich, and G. R. G. Lough. 1985. The biogeochemistry of toluene in coastal seawater: radiotracer experiments in controlled ecosystems. *Biogeochemistry* **1**: 307-328.

Wakeham, S. G., E. A. Canuel, and P. H. Doering. 1986a. Behavior of aliphatic hydrocarbons in coastal seawater: mesocosm experiments with [^{14}C]-octadecane and [^{14}C]-decane. *Environ. Sci. Technol.* **20**: 574-580.

Wakeham, S. G., E. A. Canuel, and P. H. Doering. 1986b. Geochemistry of volatile organic compounds in seawater: mesocosm experiments with ^{14}C-model compounds. *Geochim. Cosmochim. Acta* **50**: 1163-1172.

10. BALTIC SEA EUTROPHICATION: A CASE STUDY USING EXPERIMENTAL ECOSYSTEMS

Sigurd Schulz

Abstract

Remarkable changes have taken place in the Baltic Sea ecosystem in the last decade. One serious problem among these alterations is the evident eutrophication. Therefore, investigations using experimental ecosystems were carried out in the Arkona Sea in the productive seasons in order to study eutrophication effects. For this purpose, the natural water with the plankton community was enriched by nutrients to about the winter level or to about double concentrations. In spring and summer, the phytoplankton responded to this event with rapid uptake of nutrients and an increase in primary production and biomass. In autumn, uptake of nutrients was also rapid but the productivity did not increase and an increase in biomass was only visible in diatoms. With higher nutrient supply, production increased and the productive season was prolonged. This fact is of importance for the function of the pelagic system in the Baltic, because the biomass and nutrients remaining after the spring bloom determine to a great extent the productivity of the whole year.

INTRODUCTION

Eutrophication is a feature of the Baltic Sea and of nearly all natural ecosystems, and it is considered with increasing concern. It has now been shown in most compartments of the Baltic ecosystem (Elmgren, 1984; Larsson et al., 1985; Schulz, 1985). Differences in the degree of eutrophication exist depending on the area under consideration. Generally, the coastal areas and the entrances to the Baltic are more threatened. The effects of the eutrophication and the sources are known in principle; however, the pathways of the nutrients into the compartments, the transfer, and the responses to the impact are, in most cases, a matter of speculation.

This paper deals with some of the results of mesocosm experiments (ÖKEX, ecological experiments) carried out under *in situ* conditions in the

Arkona Sea. Natural (upwelling events) or artificial (man-made inputs) impacts on the pelagic system were simulated by nutrient additions, and the response of the phytoplankton was studied. An attempt is made to apply some findings of the experiments on the pelagic system and to describe possible consequences.

MATERIALS AND METHODS

The experiments were carried out at the GDR-station 113 (also a site in the frame of the Baltic Monitoring Programme of the HELCOM) at 54° 55.5' N, 13° 30.0' E. The cylindrical, 25.4-l volume enclosures floated in 2.5 m water depth, and were fastened to a 200-m long rope between two moored marker buoys. The enclosures had a top and underside made of plexiglass and a wall of 2 mm polyethylene foil. Further details on the containers and the performance of the experiments are described by Schulz et al. (1985). The experiments were carried out at the following times:

ÖKEX 81: 8 - 28 April, 1981
ÖKEX 78: 7 - 27 July, 1978
ÖKEX 83: 13 - 30 September, 1983

The chemical and biological methods used followed the standard procedures accepted for the Baltic Sea (Rohde and Nehring, 1979; Dybern et al., 1976). The doubling times of the phytoplankton (cell number) were calculated according to Knoechel and Kalff (1978) using the initial and end values.

THE BALTIC SEA ECOSYSTEM - A BRIEF OCEANOGRAPHIC INTRODUCTION

The Baltic Sea has existed in the present state for about 5,000 years and is therefore geologically and biologically a very young ecosystem. The semi-enclosed, so-called "Northern Mediterranean Sea" consists of a series of basins which are separated from each other by sills as shown in Fig. 1. The entrances to the Baltic are narrow and shallow and therefore limit water exchange with the North Sea. In connection with the positive water balance, this results in the establishment of a permanent halocline and a decrease in salinity from west to east, as well as from the bottom to the surface. The permanent halocline prevents thorough convective mixing throughout the year, and the water above is exchanged only in autumn and spring. This strongly influences the oxygen and nutrient regime of the deep water. This water tends to stagnation, oxygen depletion, and hydrogen sulfide production, and is only renewed by episodic saltwater influxes.

The Baltic is mainly inhabited by euryhaline marine organisms. Salinity is the major factor determining the distribution of the organisms. Species

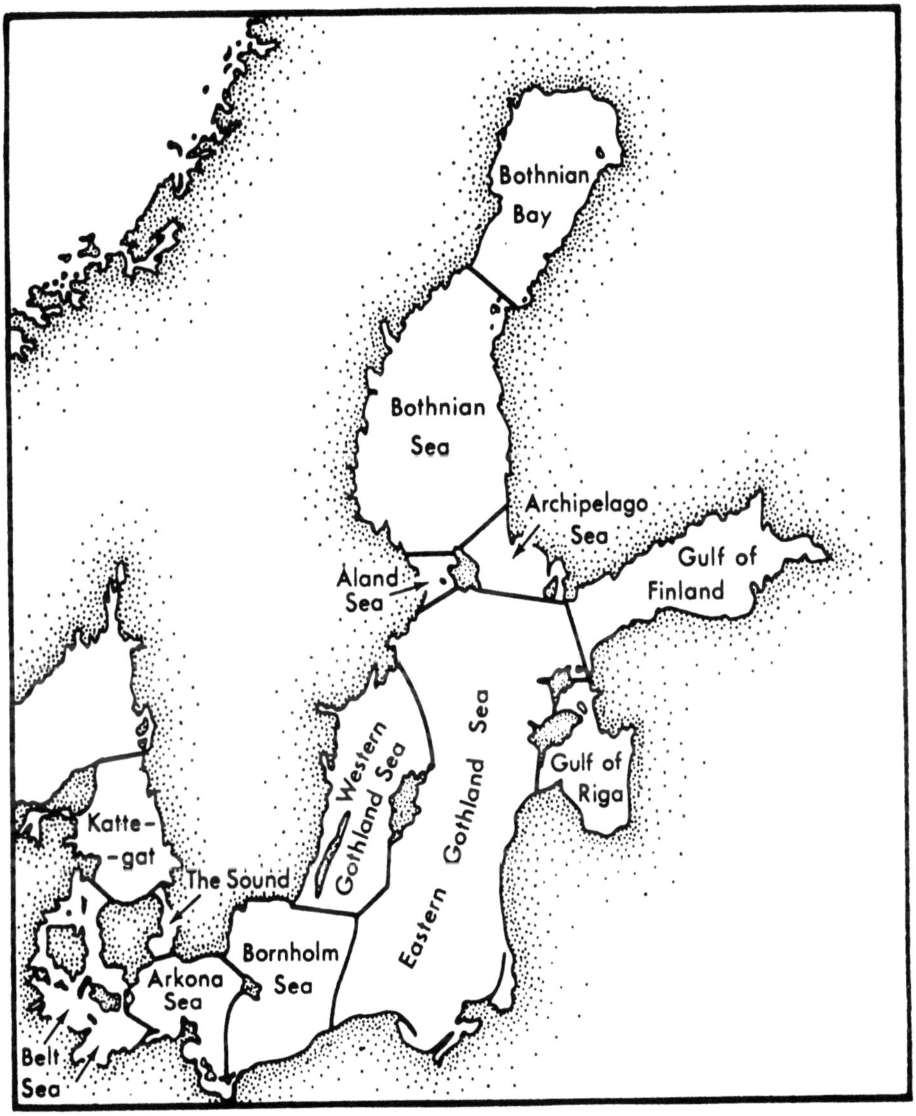

Figure 1. Map of the Baltic Sea. After Elmgren (1984).

diversity is low compared with other marine areas. Because of the highly variable environment, the biota lives under continuous stress conditions. This results in a reduction in size of specimens and in reduced productivity of the community.

The Baltic Sea is a highly variable ecosystem in which changes of important variables appear in different time and space scales. In the following discussion, some long-term changes are described. In the past decade, the temperature of the deep water increased in the different subregions by 0.6 to 1.8°C. The salinity rose in the range of 0.8 to 1.5 x 10^{-3}, and the oxygen content dropped by 2-4 cm^3 dm^{-3} (Matthäus, 1983). Similar trends could be observed in the surface water. Since 1977, no important saltwater influx has occurred to ventilate the deep basins.

Since the end of the 1960s, a 2 to 3-fold increase in phosphate and inorganic nitrogen compounds has been detected. The winter concentrations are presently 0.6 to 0.7 $\mu M\ l^{-1}$ for phosphate and between 4 to 5 $\mu M\ l^{-1}$ for nitrate (Nehring et al., 1984). The causes are seen in a continuous and also increasing input of domestic and industrial sewage, as well as nutrients from agriculture via air and water runoff (Anon., 1986). The reported changes in the physical and chemical environment also caused changes in the biota. With the increase in salinity (long-term or episodic), marine organisms invaded the Baltic (Segerstrale, 1957); the term "oceanization" was coined for the process. The increase in nutrients resulted in an expected rise in biological productivity. Increasing trends are now evident for nearly all compartments of the ecosystem from phytoplankton to pelagic fish (Schulz, 1985, 1988). The increase rates for the Arkona Sea are given in Table 1 as an example; these were calculated by means of a regression analysis. Similar values could also be obtained for other regions of the Baltic.

Table 1.

Averages of the period 1975-1983 for different parameters and the calculated annual increase.

Parameter	Average x	Mean annual increase b	Percentage b/x(100)
$NO_3 + NH_4$	4.6 $\mu M\ l^{-1}$	0.19 $\mu M\ l^{-1}$	4
Chlorophyll *a*	45.0 mg m^{-2}	3.2 mg m^{-2}	7
Primary productivity	0.76 gC $m^{-2}\ d^{-1}$	0.05 gC $m^{-2}\ d^{-1}$	6
Zooplankton biomass	0.66 ml m^{-3}	0.02 ml m^{-3}	3

RESULTS OF THE EXPERIMENTAL ECOSYSTEM APPROACHES

Spring

The experiments started just before the onset of the spring bloom. Therefore, in the first experiment (Fig. 2), the nitrate concentration showed the high winter level. This high concentration was present in the second experiment also. After about one week, the surface water was exhausted of nitrogenous nutrients, and the experiments with additions of nutrients were started (III + IV). The nitrate was taken up very rapidly. There were no differences between the four experiments.

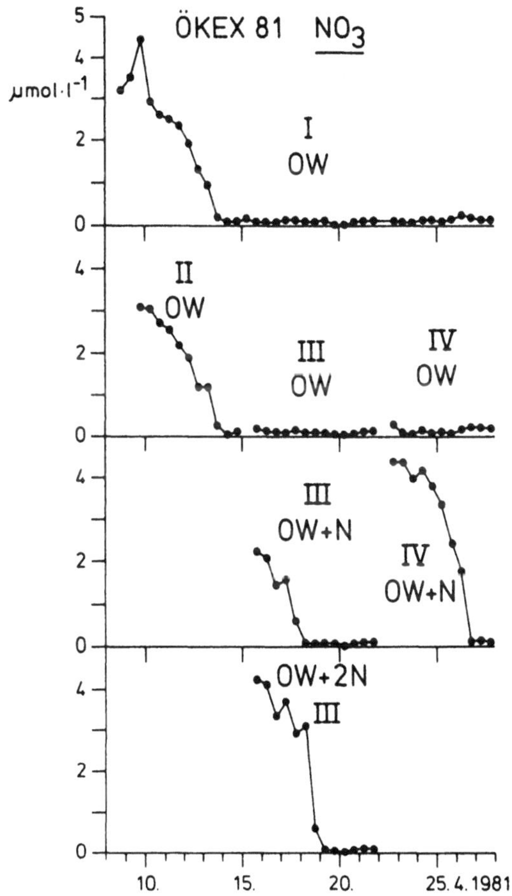

Figure 2. The nitrate concentration (μM l^{-1}) in experiments I-IV in the spring, 1981. OW, Surface water; OW + N, surface water plus nutrients; OW + 2N, surface water with doubled nutrient addition.

The rapid development of the phytoplankton biomass and the partitioning between the different groups are shown in Fig. 3. The phytoplankton was dominated by the diatoms *Thalassiosira rotula, Achnanthes taeniata, Skeletonema costatum, and Chaetoceros* spp. The numbers of microalgae were relatively small, although they also increased during the experiments. Dinoflagellates appeared only in the last two experiments.

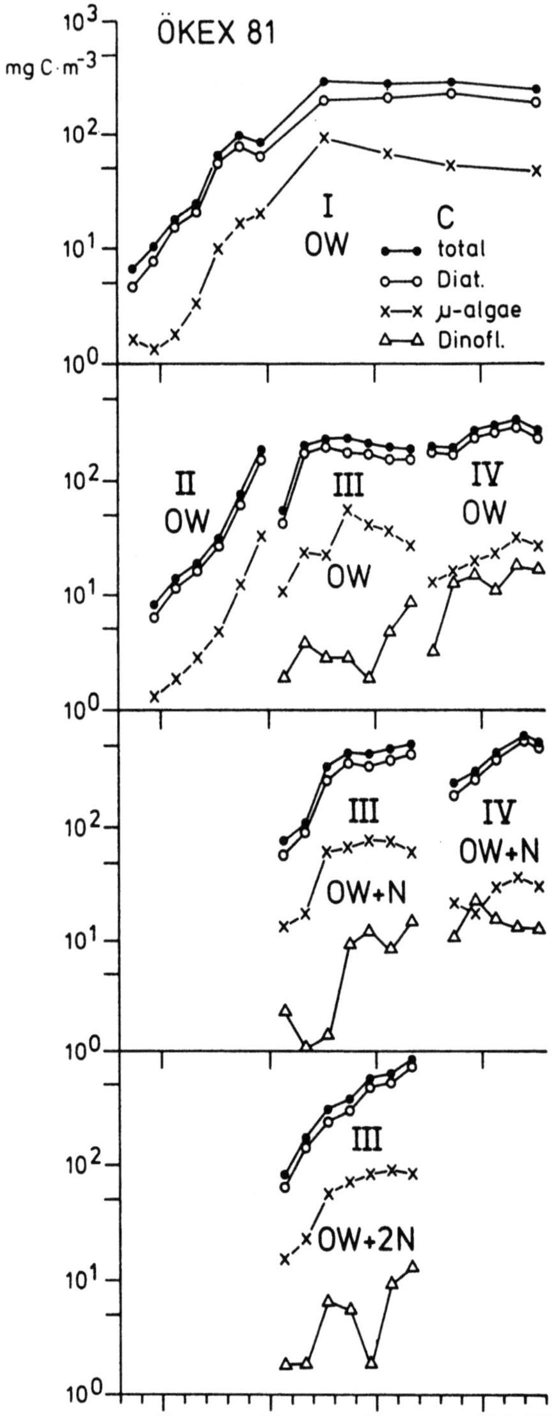

Figure 3. The phytoplankton biomass (mgC m^{-3}) during experiments I - IV in the spring, 1981. Carbon is depicted on a log scale. Legend as in Fig. 2.

Summer

The surface water in summer was completely exhausted of nitrogenous nutrients, as is typical for this season (Figs. 4 and 5). In this experiment, deep water from the station (sampling depth, 43 m) was added to the surface water in a ratio of 2:1 to simulate high ammonia input in preference to nitrate. This comes closer to the natural conditions occurring during this season in the sea area under consideration.

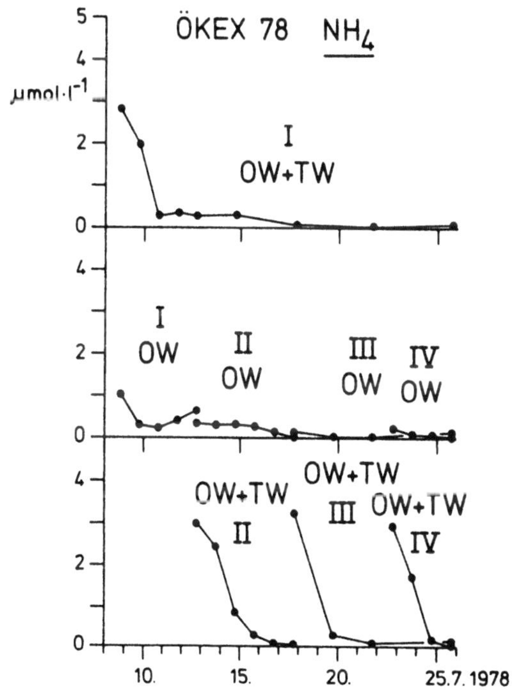

Figure 4. The concentration of ammonia (μM l^{-1}) during experiments I-IV in the summer, 1978. OW, Surface water; OW + TW, surface water plus deep water.

The phytoplankton consisted mainly of cyanobacteria and microalgae. Dinoflagellates and diatoms became more numerous during the experiments (Fig. 6). As could be seen from all experiments with deep-water additions, the microalgae responded quickly in the beginning of the experiments. Similarly, the diatoms reacted with an increase in biomass. After a few days, however, the increase in biomass was only produced by the cyanobacteria, pointing to their ability to fix molecular nitrogen as well as to other effects (e.g. recycling) (Hübel and Hübel, 1980). For the same reasons, the control bags also showed an increase in biomass, in contrast to the spring experiments.

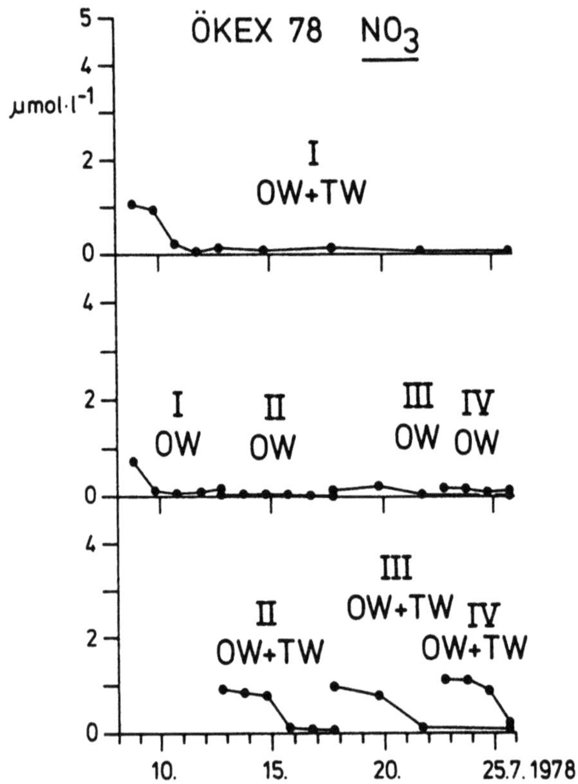

Figure 5. The nitrate concentration ($\mu M\ l^{-1}$) during experiments I-IV in the summer, 1978. Legend as in Fig. 4.

Autumn

Although the phosphate concentrations in the surface water showed a slight increase, the limiting nitrogenous compounds became nearly depleted in this layer. After additions of nutrients to about the level of the winter concentrations (Figs. 7 and 8), both ammonia and nitrate were taken up in the third experiment. In the second experiment, ammonia was preferentially used because of its high concentration.

The phytoplankton was dominated by *Rhodomonas minuta*; diatoms and dinoflagellates played a minor role, and cyanobacteria were completely absent because of the season. As can be seen from Fig. 9, the nutrient addition had nearly no effect on the total phytoplankton biomass. This was caused mainly by the inactivity of the dominating microalgae. On the other hand, diatoms and dinoflagellates showed a remarkable increase in cell number. Diatoms like *Rhizosolenia fragillissima, Coscinodiscus radiatus* and, later, *Skeletonema costatum*, as well as small dinoflagellates, became more abundant.

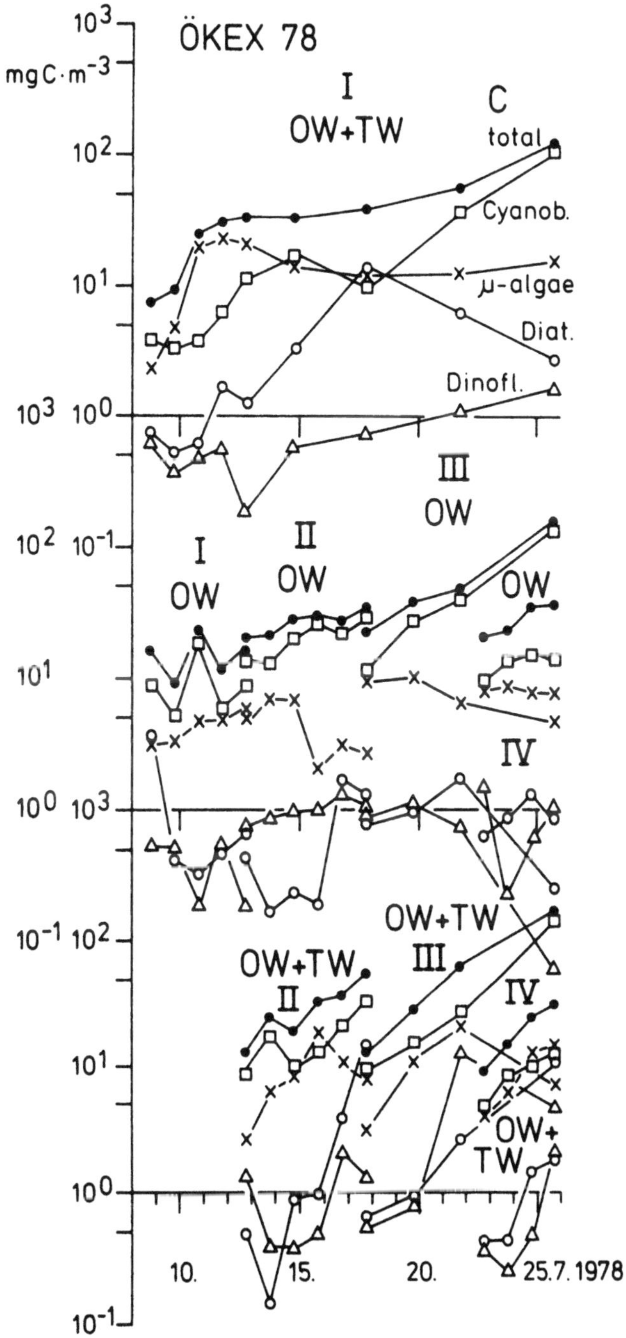

Figure 6. The phytoplankton biomass (mgC m^{-3}) during experiments I-IV in the summer, 1978. The carbon scales are overlapping, and carbon is depicted on a log scale. Legend as in Fig. 4.

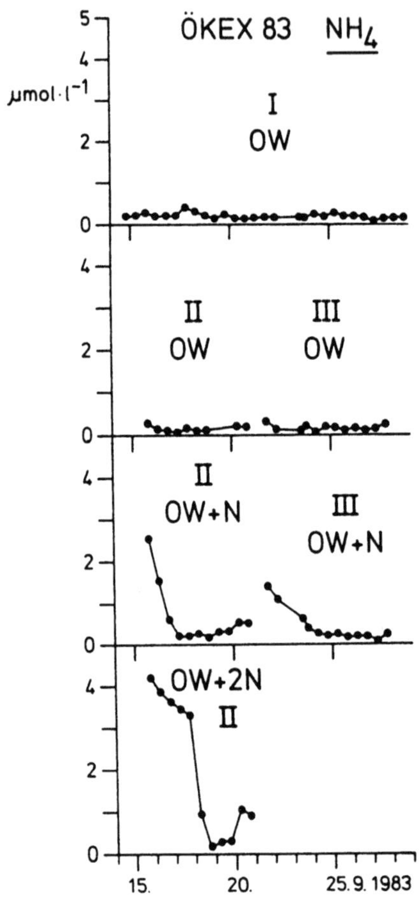

Figure 7. The concentration of ammonia ($\mu M\ l^{-1}$) during experiments I - III in the autumn, 1983. Legend as in Fig. 2.

DISCUSSION

Experiments with enclosed plankton communities ought to be considered as approaches to the natural conditions for several reasons (Boyd, 1981). They can only describe the processes which are governing natural ecosystems within certain limits. On the other hand, they offer the only possibility of studying these processes in conditions as close to the environment as possible. Apart from the restrictions of using such enclosures, short-term experiments (hours to days) were used to explain reactions of the plankton community within the long-term frame of the eutrophication process (years to decades). The results obtained, however, have been checked by findings from time-series investigations and long-term studies of the pelagic system (Nehring et al., 1984; Schulz, 1985) and were found to prove and to support them (see discussion of increased phytoplankton productivity and biomass after nutrient addition, and reactions in phytoplankton species composition).

In the experiments, of course, nutrient concentrations have been used which are much higher than the annual increase obtained from linear regression. Nehring (1981) calculated values of 0.05 μM l^{-1} per year for phosphate and 0.19 μM l^{-1} yr^{-1} for nitrate for the Arkona Sea. These concentrations could not be used in the experiments because the response of the plankton community would range within the uncertainties of the methods. Therefore, concentrations in the dimensions of the winter surface values were usually used. Also, these concentrations might indeed be maximum inputs after upwelling events that force up nutrient-rich deep water from below the halocline.

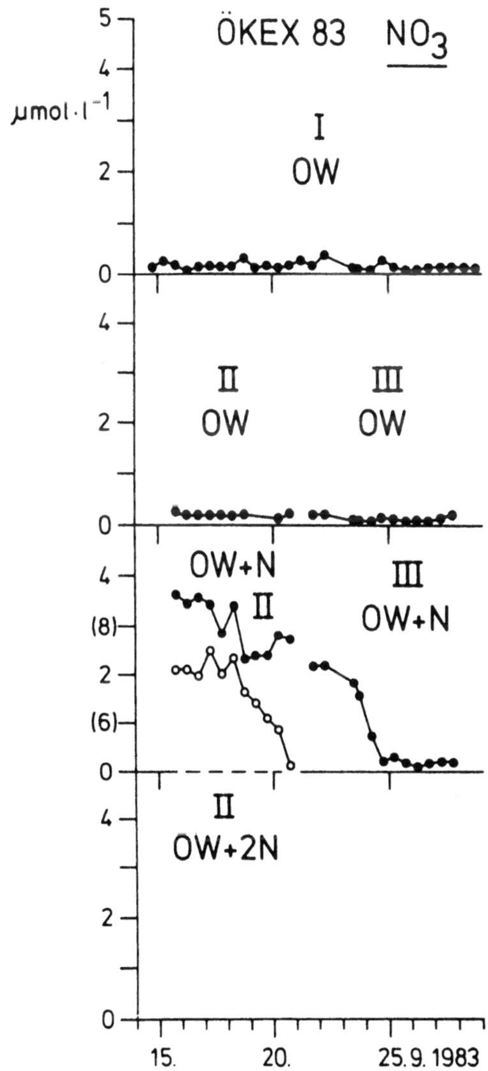

Figure 8. The nitrate concentration (μM l^{-1}) during experiments I - III in the autumn, 1983. The curve with open circles refers to experiment OW + 2N. The scale is overlapping. Legend as in Fig. 2.

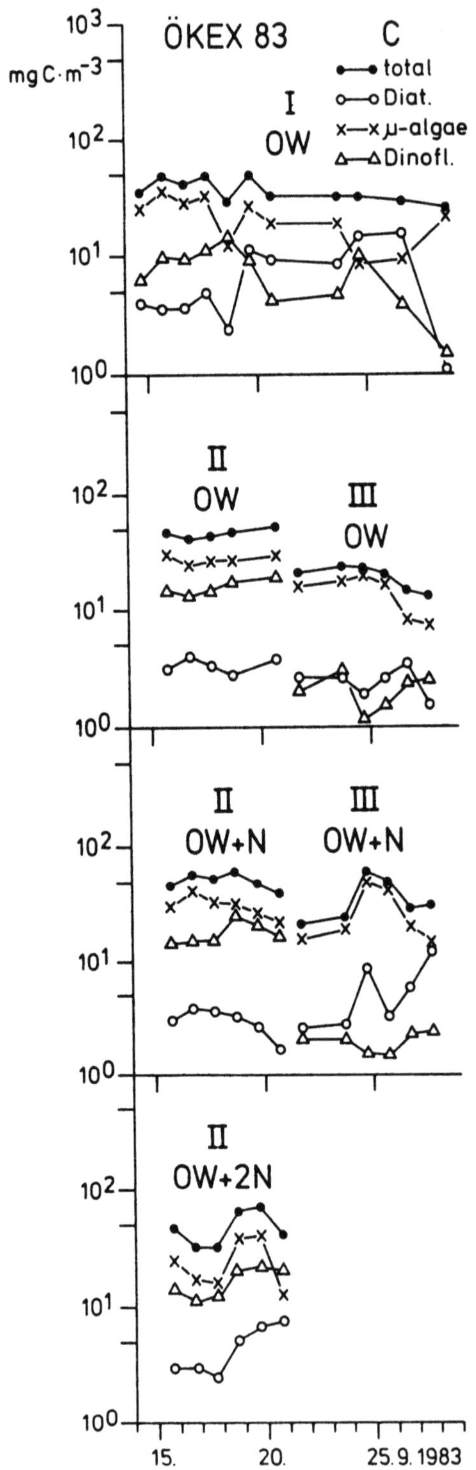

Figure 9. The phytoplankton biomass (mgC m^{-3}) during experiments I-III in the autumn, 1983. Carbon is depicted on a log scale. Legend as in Fig. 2.

Response of Phytoplankton to Nutrient Additions

Nutrient Uptake

The nitrogen compounds were taken up by the phytoplankton in the order of ammonia to nitrate (Schulz et al., 1985). This was earlier stated by Dugdale and Goering (1967) and also proved by Graneli (1981) in her experiments on the Baltic Sea phytoplankton. Both ammonia and nitrate were taken up after four days in the experiments with simulation of the winter concentrations. A doubled amount of nitrogen resulted in a prolonged uptake of about one day. This fact is especially important if we consider the proven nutrient increase in the winter surface water during the last years (Nehring, 1981), and its possible consequences are discussed later in detail.

The relative uptake of nutrients was surprisingly similar (except nitrate in autumn) in all three investigated seasons (Table 2), although important factors influencing this process (such as phytoplankton composition and physiological

Table 2.

Initial concentrations I (in $\mu M\ l^{-1}$), mean daily uptake rates U (in $\mu M\ l^{-1}\ d^{-1}$) and relative uptake of phosphate, nitrate, and ammonia in the experiments with a single nutrient addition. (n = number of values)

Nutrients	I	U	U/I(100)	n
Spring, 1981				
PO_4^{3-}	0.92	0.16	17.4	20
NO_3^-	3.20	0.65	20.3	20
NH_4^+	2.98	0.65	21.8	20
Summer, 1978				
PO_4^{3-}	0.54	0.08	14.8	20
NO_3^-	1.01	0.23	22.8	20
NH_4^+	3.09	0.55	18.1	20
Autumn, 1983				
PO_4^{3-}	0.93	0.13	14.0	15
NO_3^-	4.32	0.42	9.7	15
NH_4^+	2.66	0.51	19.2	15

stage, and light) were rather different. For phosphate at least, the dependency of the absolute uptake rates on the initial concentrations was obvious (Ketchum, 1939). This relation was not clear in the case of ammonia and nitrate , probably due to the changing relation between the two nitrogen compounds in the experiments.

Phytoplankton Biomass and Species Composition

Doubling times were used to assess changes in phytoplankton after nutrient additions. These values give more detailed information on alterations at the level of species or groups of phytoplankton than the values of chlorophyll or primary production which also were available.

As can be seen from Table 3, which shows two consecutive periods from experiment I (compare Fig. 2), the doubling activity of the diatoms and microalgae decreased with the depletion of the nutrient stock. The response varied with different species, however. Table 4 shows a phase when the nutrients were exhausted and additions were performed in two different concentrations. An effect of the nutrient addition was apparent, as well as the concentration dependency at least in this level. Striking differences existed between the species.

Table 3.

Doubling time (t_D in hours) of phytoplankton species groups (cell number) during experiment I in spring, 1981.
(SW, surface water; WCN, whole cell number)

	SW (8. to 13.4.1981)	SW (14. to 20.4.1981)
WCN	34.4	80.8
Microalgae	35.7	81.6
Thalassiosira spp.	28.0	268.5
Skeletonema costatum	34.0	91.3
Chaetoceros spp.	23.4	79.6

Table 5 exemplifies the situation during the summer experiment. In this case also, an effect on the phytoplankton activity could be shown in all groups; however, the different groups exhibited remarkably different responses. The slow reactions of microalgae and dinoflagellates were in sharp contrast to the high activity in diatoms and in cyanobacteria.

Table 4.

Doubling time (t_D in hours) of phytoplankton species groups (cell number) during experiment II (15. to 21.4) in spring, 1981. (SW, surface water; SW + N, surface water plus nutrients; SW + 2N, surface water plus doubled amount of nutrients; WCN, whole cell number)

	SW	SW + N	SW + 2N
WCN	102.1	49.7	42.1
Microalgae	111.1	65.5	57.6
Thalassiosira spp.	82.1	74.6	61.3
Skeletonema costatum	59.1	46.3	39.2
Chaetoceros spp.	126.4	33.9	28.5

The autumn conditions reflected the low activity of the dominating flagellate stock and, therefore, the unimportant changes in the total biomass. Diatoms and dinoflagellates, on the other hand, responded rapidly and showed a clear decrease in the doubling times (Table 6).

In no case did alterations occur in the species composition of the dominating forms during the three seasons of the experiments. However, changes in the dominance relations did appear because of the differing responses of the considered groups to the nutrient additions.

Table 5.

Doubling time (t_D in hours) of phytoplankton species groups (cell number) during experiment III (17. to 25.7) in summer, 1978. (SW, surface water; SW + DW, surface water plus deep water; WCN, whole cell number)

	SW	SW + DW
WCN	-226.5	140.6
Microalgae	-186.0	158.0
Cyanobacteria	82.8	68.3
Dinoflagellates	- 53.8	534.3
Diatoms	50.1	32.5

Table 6.

Doubling time (t_D in hours) of phytoplankton species groups (cell number) during experiment III (21. to 27.9) in autumn, 1983. (SW, surface water; SW + N, surface water plus nutrients; WCN, whole cell numbers)

	SW	SW + N
WCN	-165.4	2613.7
Microalgae	-164.8	2126.6
Dinoflagellates	64.0	48.7
Diatoms	228.6	83.1

As could be concluded from the response of the phytoplankton to nutrient addition in the three seasons, the clearest reaction was noticed in spring. In summer and autumn, only a part of the phytoplankton community responded to the impact. Because the cyanobacteria and microalgae dominated the biomass as well as the primary production in summer, the chlorophyll content showed an increase. In autumn, however, the dominating microalgae did not respond; only the small numbers of diatoms and dinoflagellates responded to nutrient addition. Therefore no reaction could be observed in the total biomass, primary production, and chlorophyll (Schulz, 1985).

However, in autumn, the phytoplankton took up the nutrients at nearly the same speed and amount as in spring and summer experiments. Possibly the high amount of added nutrients was a physiological stress for the low-nutrient-adapted microalgae (Parsons et al., 1977). Another explanation could be that starving phytoplankton have anomalous C:N and C:P ratios with a surplus of carbohydrates and lipids (Healey, 1979). After nutrient addition, absorbed nutrients will be used for the synthesis of stored carbon to protein. There is no need for photosynthesis in this phase. Lean and Pick (1981) obtained similar results and concluded that, in the case of a very low internal nutrient pool, primary productivity measurements are insufficient as an activity measure.

Possible Consequences of Nutrient Additions on the Pelagic Ecosystem of the Baltic Sea

As the experiments have shown, the input of nutrients leads to higher phytoplankton activity (primary production, biomass, and chlorophyll) and also to a prolongation of the productive period (Schulz et al., 1985). This seems to be especially important in spring because then the gap in the time scales of phyto- and zooplankton development is lessened. This means that a larger share of the spring phytoplankton biomass might be used by the developing zooplankton, thus increasing their biomass and retaining a larger part of chemical energy in the pelagic system. In this way the summer production might be influenced positively.

The experimental results, however, also show that the input of nutrients in different seasons does not lead to changes in the species composition. Only the dominance relations change as a result. This is also the result of most prior phytoplankton studies which have concluded that, although the winter nutrient level may be increased significantly, changes in species composition could not be shown (Wulff et al., 1986).

Nutrient inputs to the pelagic system during different seasons have different meanings for the ecosystem. In spring, the phytoplankton production responds fastest to the impact and, because of the early stage of the zooplankton development, the largest part of the addition is sedimented and acts as an oxygen sink in deep water. Only a small share increases the productivity over the whole year, as was mentioned above (Schulz, 1985).

The summer situation in the Baltic is characterized by a well balanced pelagic community (Schulz and Breuel, 1984), with a low, grazed-down, phytoplankton stock and a highly developed zooplankton community. This system also reacts very rapidly to an input of nutrients. In contrast to the spring conditions, however, the additional energy leads to an enhancement of the productivity in all compartments of the pelagic system. The organic substance is retained in the euphotic layer and circulated several times; sedimentation is negligible during this time (Prandke, 1980; Peinert et al., 1982). If the additional production is not completely transferred to fish biomass or to higher predators, then it can later threaten the oxygen supply in the deep water again.

The autumn situation is similar to the spring condition because, again by the decrease in the zooplankton stock, the pelagic system tends to disequilibrium leading to a higher sedimentation of newly produced algal biomass.

LITERATURE CITED

Anonymous. 1986. First periodic assessment of the state of the marine environment of the Baltic Sea area, 1980-1985. General conclusions. *Baltic Sea Environment Proceedings*. No. 17A. 56 pp.

Boyd, C. M. 1981. Microcosms and experimental planktonic food chains. Pp. 627-650. **In**: A. R. Longhurst [ed.], *Analysis of Marine Ecosystems.* London: Academic Press.

Dugdale, R. C. and I. J. Goering. 1967. Uptake of new regenerated forms of nitrogen in primary productivity. *Limnol. Oceanogr.* **12**: 685-695.

Dybern, B. I. , H. Ackefors, and R. Elmgren [eds.]. 1976. *Recommendations on Methods for Marine Biological Studies in the Baltic Sea.* BMB Publ. No. 1. 98 pp.

Elmgren, R. 1984. Trophic dynamics in the enclosed brackish Baltic Sea. *Rapp. P.-V. Réun. Cons. Perm. Int. Explor. Mer* **183**: 152-169.

Graneli, E. 1981. Experimental investigations of limiting nutrients for phytoplankton production in the brackish-water Öresund SW Sweden. Lund: Unpubl. thesis. Pp. 1-18.

Healey, F. P. 1979. Short-term responses of nutrient deficient algae to nutrient addition. *J. Phycol.* **15**:189-299.

Hübel, H. and M. Hübel. 1980. Nitrogen fixation during blooms of *Nodularia* in coastal waters and back waters of Arkona Sea (Baltic Sea) in 1974. *Int. Rev. Ges. Hydrobiol.* **65**: 793-808.

Ketchum, B. H. 1939. The absorption of phosphate and nitrate by illuminated cultures of *Nitzschia closterium. Am. J. Botany* **26**: 399-407.

Knoechel, R. and J. Kalff. 1978. An *in situ* study of the productivity and population dynamics of five freshwater planktonic diatom species. *Limnol. Oceanogr.* **23**: 195-218.

Larsson, U., R. Elmgren, and F. Wulff. 1985. Eutrophication and the Baltic Sea: causes and consequences. *Ambio* **14**: 9-14.

Lean, D. R. S. and F. R. Pick. 1981. Photosynthesis response of lake plankton to nutrient enrichment. A test for nutrient limitation. *Limnol. Oceanogr.* **26**: 1001-1019.

Matthäus, W. 1983. Aktuelle Trends in der Entwicklung des Temperatur-, Salzgehalts- und Sauerstoffregimes im Tiefenwasser der Ostsee. *Beitr. Meereskunde* **49**: 47-64.

Nehring, D. 1981. Hydrographisch-chemische Untersuchungen in der Ostsee von 1969-1978. *Geod. Geoph. Veröff.* R. IV. **35**: 39-220.

Nehring, D., S. Schulz, and W. Kaiser. 1984. Long-term phosphate and nitrate trends in the Baltic proper and some biological consequences: A contribution to the discussion concerning the eutrophication of these waters. *Rapp. P.-V. Réun. Cons. Perm. Int. Explor. Mer* **183**: 193-203.

Parsons, T. R., K. von Bröckel, P. Koeller, M. Takahashi, M. R. Reeve, and O. Holm-Hansen. 1977. The distribution of organic carbon in a marine planktonic food web following nutrient enrichment. *J. Exp. Mar. Biol. Ecol.* **26**: 235-247.

Peinert, R., A. Saure, P. Stegmann, C. Stienen, H. Haardt, and V. Smetacek. 1982. Dynamics of primary production and sedimentation in a coastal ecosystem. *Neth. J. Sea Res.* **16**: 176-289.

Prandke, H. 1980. Einige Ergebnisse der Lichtstreuuntersuchungen in der mittleren Ostsee. *Beitr. Meereskunde* **44/45**: 43-54.

Rohde, K. H. and D. Nehring. 1979. Ausgewählte Methoden zur Bestimmung von Inhaltsstoffen im Meer- und Brackwasser. *Geod. Geoph. Veröff.,* R. IV. **27**: 1-68.

Schulz, S. 1985. Ergebnisse ökologischer Untersuchungen im pelagischen ökosystem der Ostsee. Als Manuskript gedruckt, Rostock - Warnemünde. 185 pp.

Schulz, S. 1988. Changes in the Baltic pelagic system. *Proc. 21st EMBS,* Gdansk. 18 pp.

Schulz, S. and G. Breuel. 1984. On the variability of some biological parameters in the summer pelagic system of the Arkona Sea. *Limnologica* **15**: 365-370.

Schulz, S., G. Breuel, A. Irmisch, H. Siegel, and V. Kell. 1985. Ergebnisse ökologischer Untersuchungen an eingeschlossenen Planktongemeinschaften der Arkonasee im Frühjahr 1981. *Geod. Geoph. Veröff.,* R. IV. **41**: 1-66.

Segerstrale, S. 1957. Baltic Sea. *Mem. Geol. Soc. Amer.* **67**: 751-800.

Wulff, F., G. Aertebjerg, G. Nicolaus, A. Niemi, P. Ciszewski, S. Schulz, and W. Kaiser. 1986. The changing pelagic ecosystem of the Baltic Sea. *Ophelia* Supp. **4**: 299-319.

11. MESOCOSMS: STATISTICAL AND EXPERIMENTAL DESIGN CONSIDERATIONS

John C. Gamble

Abstract

Large-scale enclosure (mesocosm) experiments do not lend themselves to conventional methods of experimental design involving the use of inferential statistics. When considering design criteria for mesocosm experiments, it is therefore necessary to examine the rigor of the proposed replication, to make allowance for inherent spatial and temporal heterogeneity, and to consider alternative experimental strategies.

INTRODUCTION

The difficulty in applying conventional inferential statistics to results from mesocosm experiments is one of the major criticisms of the enclosure technique. However, such limitations, while real and in some cases unavoidable, should not inevitably mitigate against the use of mesocosms but should be realized and countered by careful procedures and by resisting the temptation to apply statistics uncritically. There are many powerful reasons why we should carry out experiments in mesocosms, and these must take precedence over the shortcomings of the technique.

My purpose in this brief review will be to draw attention to the difficulties in applying statistical techniques to mesocosm data, to discuss pertinent papers, and to make some recommendations.

MESOCOSM DESIGNS: CONCEPTUAL AND PRACTICAL

There are many good reasons for experimenting with large and complex systems (Menzel and Steele, 1978). For instance, complex enclosed systems can be experimentally manipulated in order to test hypotheses about ecosystem mechanisms. They have also been used extensively and most effectively for deducing broad-scale effects of pollutants. It is, however, this very complexity which has caused problems with experimental design and with the application of rigorous statistical procedures, particularly with regard to the question of replicability. Further, mesocosm experiments *sensu strictu* are long term, frequently extending for periods of weeks during which interactions between trophic levels can become evident. In such circumstances the populations within individual mesocosms, which initially are set up as replicates, frequently diverge from one another even though the physico-chemical conditions are practically identical.

It is necessary to recognize the variety of experimental systems currently in use under the umbrella term "mesocosm." Such systems range from simple plastic film or mesh bags of less than 5 m^3 volume to pond systems of several thousand m^3 capacity. Some mesocosms isolate the water column from the benthos, some have included both ecosystems (especially in freshwater systems), and there is now a growing emphasis on benthic enclosures. Hence, in a statistical sense, it is impossible to make specific recommendations for mesocosms. Their variety demands that each system be treated individually.

APPLICATION OF STATISTICAL METHODS

As in all scientific practices, statistical techniques are applied for two reasons: objective description and verifiable inference. The former is concerned with making measurements of environmental parameters which are a true representation of reality, the latter with the interpretation of experimental manipulations.

Descriptive Statistics

In many senses it is easier to handle the description of events in mesocosms than in natural field situations. (It is one of the reasons for doing mesocosm experiments in the first place!) In pelagic systems, the mesocosm largely overcomes the problem of advection and the need to resort to Lagrangian sampling strategies to carry out effective time-series analyses. However, as pointed out by Lawson and Grice (1977), Stephenson et al. (1984) and de Lafontaine and Leggett (1987a), populations of pelagic organisms are not distributed uniformly within the systems, either horizontally or vertically, and similar heterogeneity must also be met in benthic systems (Underwood and Peterson, 1988; Warwick et al., 1988).

Therefore, it is essential to design sampling protocols which take account of spatial and temporal heterogeneity in the systems and which provide an estimate of sampling error. As always, this is a compromise between idealism and realism. It is relatively easy to make replicate measurements of chemical parameters such as nutrient levels, more difficult to make rate measurements such as primary production, and usually totally impractical even to duplicate the detailed analyses of phyto- and zooplankton population samples. The mesocosm sampling philosophy is best summarized by the concluding remarks of Stephenson et al. (1984): "Our studies have identified a potential problem that all investigators using enclosures for toxicology studies must resolve. The distribution of organisms within enclosed water columns should be investigated prior to experimentation, and sampling strategies adapted to reduce error in estimates of population density." It should also be emphasized that operators need to be aware of changes in distribution with time as well as in space.

Inferential Statistics

The extreme restrictions, and in some cases the inability to apply inferential techniques to mesocosm experiments, have invalidated the mesocosm approach in the minds of many critics. Such techniques (e.g. ANOVA) were developed for experimental systems where all variables could be controlled or accounted for and where multiple replication of systems was the accepted norm. Unfortunately, the concept of the mesocosm as an experimental system which is comprised of several interacting trophic levels does not fit the strict statistical definition of a control, nor is it practical to replicate such large systems in sufficient quantity. While accepting these limitations, we must consider whether biological results from mesocosm experiments are "soft" (*sensu* Hurlbert, 1984) merely because conventional statistical evaluations cannot be really applied to them satisfactorily.

Several authors have discussed the replicability of mesocosm data (e.g. Takahashi et al., 1975; Menzel and Case, 1977; Menzel, 1977; Pilson et al., 1979; de Lafontaine and Leggett, 1987a), both in relation to how well they reproduce the external situation and how well they replicate internally. However, the complexity of multitrophic mesocosms and the methods used to initiate such experiments (e.g. by capturing a water column) in combination with the observation that, once enclosed, biological characteristics of parallel systems show divergence during the experimental time course, led Hurlbert (1984) to criticize the conclusions of the CEPEX evaluations. He concluded that, from a strictly statistical viewpoint, large multitrophic mesocosms cannot possibly replicate. Further, the mesocosm operator has no control over events during an experiment other than the application of known physico-chemical or biological entities. Thus, for large pelagic systems in particular, it is difficult to meet the ideal defined by Underwood and Peterson (1988): "In a properly designed experiment using mesocosms, the application of pollutants can be controlled so that causation can be identified without confounding due to other sources of variation between polluted and control samples." In fact, I would predict that the degree of natural variation within both treated and control mesocosms is such that subtle differences between them induced by added pollutants will be extremely difficult to verify statistically.

EXPERIMENTAL STRATEGY FOR MESOCOSMS

In the face of criticism it is necessary to develop an accepted experimental procedure for mesocosm research. If there is a need to use inferential methods such as ANOVA, rigorously replicated control and treatment systems must be deployed which of necessity will reduce the size and should increase the simplicity of the systems. Conceptually this could stretch the mesocosm technique such that the only mesocosmic attribute is the size of the systems themselves. Recent investigations of larval fish predator-prey relationships by Gamble and Fuiman (1987), de Lafontaine and Leggett (1987b), and Fuiman and Gamble (1988) exemplify this reductionism in the quest for adequate control and replication. Strictly speaking such experimental systems are merely very large enclosures, as they do not set out to contain several trophic levels of an ecosystem in their natural proportions (Menzel and Steele, 1978).

The MERL mesocosms are probably the most extensively used of all systems and have also been subject to considerable statistical investigation for the past ten years or so. In particular, emphasis was placed on the use of multivariate methods (Oviatt et al., 1977, 1980) in an attempt to describe ecosystem behavior in a holistic sense rather than in terms of comparing single variates. The exercise was reasonably successful but, as with many multivariate applications, was difficult to interpret. Oviatt et al. (1980) wondered "whether the techniques we have used are more useful than visually interpreting plots of the response of the individual variables over time." Interestingly, in one of their most recent papers, Oviatt et al. (1987) describe a nine-unit experiment with three replicate controls, but have made no pretence to any inferential statistics. The observed differences spoke for themselves.

How therefore can inferential statistics be applied to such true mesocosms? A way around the complex sequential variability of time-series data, such as population levels of individual species, is to smooth the data using a running average technique as has often been employed for modelling purposes (Hay et al., 1988) or to use some means of reducing a complex data set to a single variable. Oviatt (pers. comm.) has had success applying ANOVA to annual or seasonal means with two or three replicates, while Smith et al. (1982) proposed the use of a dissimilarity index for comparing pollutant effects on both zooplanktonic and benthonic population data. This latter procedure might satisfy statistical criteria, but it is grossly reductionist and to a certain extent defeats the objective of the mesocosm approach.

It could be argued that it is impossible to design replicated Fisherian control-treatment experiments for complex mesocosms. There are, however, two other designs that could be more acceptable: the time-series approach and the gradient experiments (Fig. 1). Smith et al. (1982) proposed that the "ABA" time-series format is highly suitable for mesocosm experiments. In this, the control condition "A" is followed by experimental period "B", returning to control period "A". Such a strategy was adopted by Parsons et al. (1986) and by Gamble et al. (1987) for dosing mesocosms with mine tailings and oilfield production water, respectively. A possible drawback when dealing with chemical pollutants, however, is that the second "A" phase is essentially a recovery phase rather than an abrupt return to control conditions. Where the perturbation is physical, such as mixing the system, cessation of effect would be immediate. In these experiments, the strongest evidence of a causal effect will be a response in the system which mirrors the treatment. Such experiments do not need controls although, where there is much covariation as in mesocosms, untreated controls are almost essential.

The other design, the gradient experiment, again dispenses with the need for replication because the effect can be described by regression analysis. In this case, treatments are graded in a series of increasing concentrations often, but not necessarily, incorporating a treatment-free, "zero", control. Parsons et al. (1977) maintained four mesocosms at four different nutrient levels and measured a graded effect on primary production, while more recently Oviatt et al. (1987) deployed three levels of sewage sludge and nutrient addition (x1, x4, and x8). The result of both series of experiments unequivocally demonstrated the effects of treatments.

CONCLUSION

Little mention has been made in this review of hypotheses testing, despite the central role of this procedure in experimental design (see Green, 1979). In truth it is impossible to reject a null hypothesis in complex mesocosm systems where true replication cannot be achieved and where the practical measures needed to attain meaningful significance levels are unattainable. Nevertheless, mesocosms are a valid ecological tool insomuch as they are closer to reality than laboratory systems and frequently are the only means of investigating effects on a multitrophic scale. Hence, the mesocosm experimenter must not look to prove an effect in the conventional sense, but must use the mesocosm as a means of recognizing trends, levels, and interrelationships which might be altered in the course of a particular treatment. In such cases the judicious use of ecological models might be more

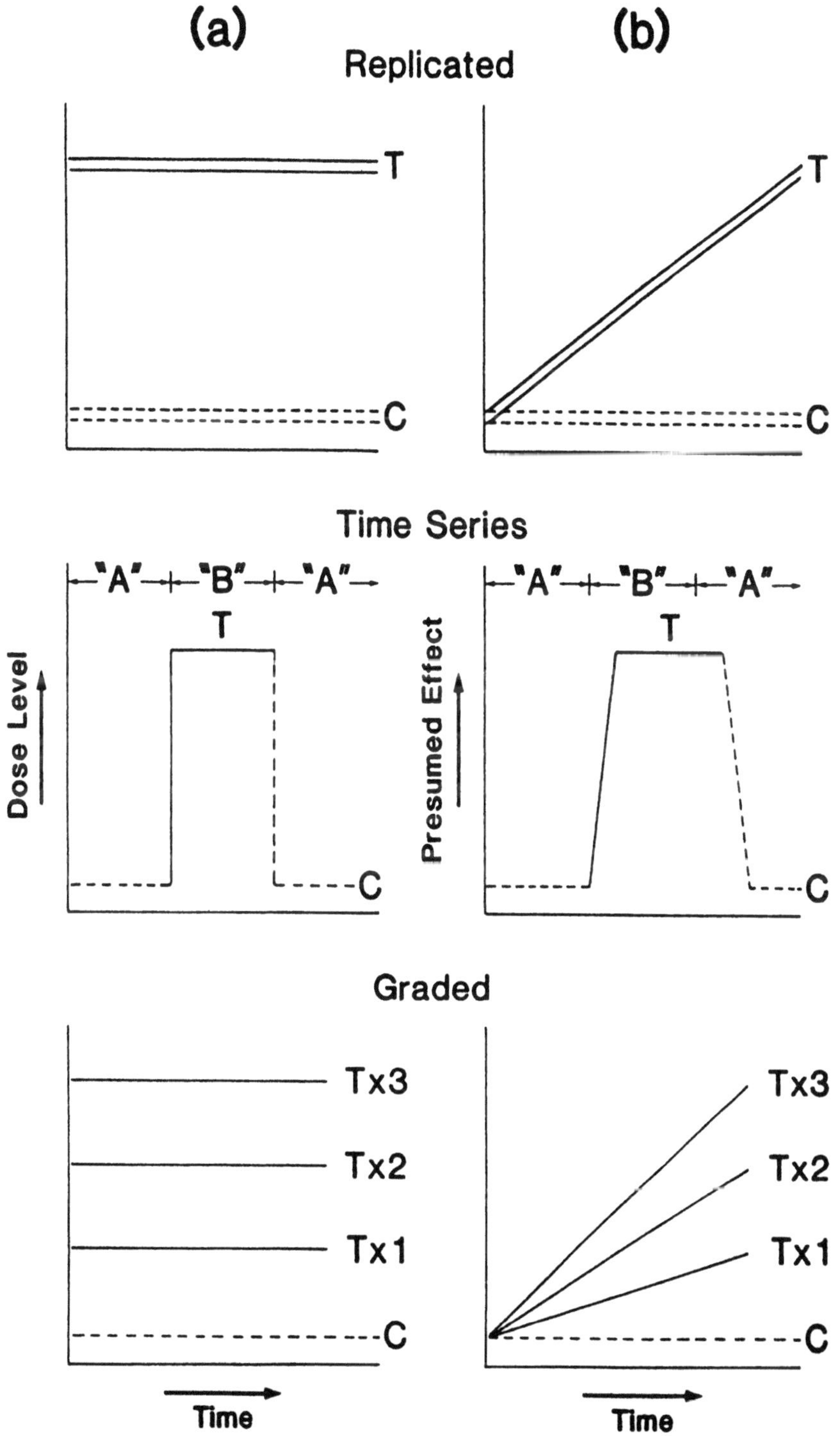

Figure 1. Representation of (a) different design protocols and (b) their presumed effects during the course of an experiment. T, treatments; C, controls.

appropriate than inferential statistics, particularly as the boundary conditions of the individual systems are so well defined.

Mesocosm experiments have been justly criticized from an inferential statistical standpoint, but this does not in itself invalidate the technique. It merely emphasizes limitations which, if recognized, should not inhibit the production of "hard" data of a unique nature.

I make three recommendations for the design of mesocosm experiments:

1. Question whether the replication proposed in an experiment is adequate for the intended purpose (Cohen, 1979; Hurlbert, 1984).

2. Adopt sampling protocols which allow for spatial and temporal heterogeneity; if possible, carry out preliminary trials and make estimates of sampling error (Stephenson et al., 1984).

3. Question whether a replicated "control-treatment" design is really necessary and look to alternatives such as time-series or gradient designs (Smith et al., 1982).

Acknowledgments

I wish to thank Torgier Bakke, Bob Clarke, George Grice, Candace Oviatt and Viktor Øiestad who all answered my plea for help when I set out on this review. My colleague Steve Hall critically reviewed and consequently much improved my initial manuscript.

LITERATURE CITED

Cohen, J. 1979. *Statistical Power Analysis for the Behavioural Sciences.* New York: Academic Press. 474 pp.

Fuiman, L. A. and J. C. Gamble. 1988. Predation by Atlantic herring, sprat and sandeels on herring larvae in large enclosures. *Mar. Ecol. Prog. Ser.* **44**: 1-6.

Gamble, J. C. and L. A. Fuiman. 1987. Evaluation of *in situ* enclosures during a study of the importance of starvation to the vulnerability of herring larvae to a piscine predator. *J. Exp. Mar. Biol. Ecol.* **113**: 91-103.

Gamble, J. C., J. M. Davies, S. J. Hay, and F. K. Dow. 1987. Mesocosm experiments on the effects of produced water discharges from offshore oil platforms in the northern North Sea. *Sarsia* **72**: 383-386.

Green, R. 1979. *Sampling Design and Statistical Methods for Environmental Biologists.* New York: Wiley. 257 pp.

Hay, S. J., G. T. Evans, and J. C. Gamble. 1988. Birth, growth and death rates for enclosed populations of calanoid copepods. *J. Plankton Res.* **10**: 431-454.

Hurlbert, S. 1984. Pseudoreplication and the design of ecological fish experiments. *Ecol. Monogr.* **54**: 187-211.

de Lafontaine, Y. and W. C. Leggett. 1987a. Evaluation of *in situ* enclosures for larval fish studies. *Can. J. Fish. Aquat. Sci.* **44**: 54-65.

de Lafontaine, Y. and W. C. Leggett. 1987b. Effect of container size on estimates of mortality and predation rates in experiments with macrozooplankton and larval fish. *Can. J. Fish. Aquat. Sci.* **44**: 1534-1543.

Lawson, T. J. and G. D. Grice. 1977. Zooplankton sampling variability: controlled ecosystem pollution experiment. *Bull. Mar. Sci.* **27**: 80-84.

Menzel, D. W. 1977. Summary of experimental results: controlled ecosystem pollution experiment. *Bull. Mar. Sci.* **27**: 142-145.

Menzel, D. W. and J. Case. 1977. Concept and design: controlled ecosystem pollution experiment. *Bull. Mar. Sci.* **27**: 1-7.

Menzel, D. W. and J. H. Steele. 1978. The application of plastic enclosures to the study of pelagic marine biota. *Rapp. P.-V. Réun. Cons. Perm. Int. Explor. Mer* **173**: 13-21.

Oviatt, C. A., K. T. Perez, and S. W. Nixon. 1977. Multivariate analysis of experimental marine ecosystems. *Helgol. Wiss. Meeresunters.* **30**: 30-46.

Oviatt, C. A., H. Walker, and M. E. Q. Pilson. 1980. An exploratory analysis of microcosm ecosystem behaviour using multivariate techniques. *Mar. Ecol. Prog. Ser.* **2**: 179-191.

Oviatt, C. A., J. G. Quinn, J. T. Maughan, J. J. Ellis, B. K. Sullivan, J. N. Gearing, P. J. Gearing, C. D. Hunt, P. A. Sampou, and J. S. Latimer. 1987. Fate and effects of sewage sludge in the coastal marine environment; a mesocosm experiment. *Mar. Ecol. Prog. Ser.* **41**: 187-203.

Parsons, T. R., K. von Brockel, P. Koeller, M. Takahashi, M. R. Reeve, and O. Holm-Hansen. 1977. The distribution of organic carbon in a marine planktonic food web following nutrient enrichment. *J. Exp. Mar. Biol. Ecol.* **26**: 235-247.

Parsons, T. R., P. Thompson, Wu Yong, C. M. Lalli, Hou Shumin, and Xu Huaishu. 1986. The effect of mine tailings on the production of plankton. *Acta Oceanologica Sinica* **5**: 417-423.

Pilson, M. E. Q., C. A. Oviatt, G. L. Vargo, and S. L. Vargo. 1979. Replicability in MERL microcosms: initial observations. Pp. 359-381. **In**: F. S. Jacoff [ed.], *Advances in Marine Environmental Research.* Proceedings of a Symposium, June 1977. EPA-600/9-79-035, US, EPA, Environ. Res. Lab. Narragansett, RI.

Smith, W., V. R. Gibson, and J. F. Grassle. 1982. Replication in controlled marine systems: presenting the evidence. Pp. 217-225. **In**: G. D. Grice and M. R. Reeve [eds.], *Marine Mesocosms. Biological and Chemical Research in Experimental Ecosystems.* New York: Springer-Verlag.

Stephenson, G. L., P. Hamilton, N. K. Kanshik, J. B. Robinson, and K. R. Solomon. 1984. Spatial distribution of plankton in enclosures of three sizes. *Can. J. Fish. Aquat. Sci.* **41**: 1048-1054.

Takahashi, M., W. H. Thomas, D. L. R. Seibert, J. Beers, P. Koeller, and T. R. Parsons. 1975. The replication of biological events in enclosed water columns. *Arch. Hydrobiol.* **76**: 5-23.

Underwood, A. J. and C. H. Peterson. 1988. Towards an ecological framework for investigating pollution. *Mar. Ecol. Prog. Ser.* **46**: 227-234.

Warwick, R. M., M. R. Carr, K. R. Clarke, J. M. Gee, and R. H. Green. 1988. A mesocosm experiment on the effects of hydrocarbon and copper pollution on a sublittoral soft-sediment meiobenthic community. *Mar. Ecol. Prog. Ser.* **46**: 181-191.

12. THE USE OF MATHEMATICAL MODELS IN CONJUNCTION WITH MESOCOSM ECOSYSTEM RESEARCH

T. R. Parsons

Abstract

The use of mathematical models in conjunction with mesocosm research has been reviewed. Models have assisted in illustrating the shortcomings of mesocosm research. At the same time, ecosystem models have profited from good time-series data such as can only be obtained from mesocosm experiments. Mathematical models should continue to be an important integral part of mesocosm research.

INTRODUCTION

Mesocosm data are ideal for testing mathematical ecosystem models. The advantage of mesocosm data over data derived from a research vessel cruise is that the former represent a continuous time-series examination of different trophic levels, whereas the latter have been subject to physical change by forces causing advection or dispersion. Mesocosms provide an excellent data bank from which to ground-truth mathematical models. In a reversal of this role, the data also assist the modeller to further develop his model from observations encountered in the mesocosm ecosystem. In the following discussion, a number of examples are given of mathematical models used in conjunction with ecosystem data. These examples range from a comparison of the accuracy of two techniques where the evaluation used was entirely statistical, to models that provide a caricature of ecosystem dynamics, and finally to those more sophisticated models that actually attempt to reproduce an ecosystem.

In this review, no attempt is made to reproduce the mathematics of ecosystem modelling; these are covered in a number of recent reviews (e.g.

Platt et al., 1981). The review is intended as an evaluation of results obtained by models as compared with actual mesocosm experiments.

RESULTS AND DISCUSSION

In the simplest case, mesocosm data collected by two techniques that are supposed to be measuring the same property may be compared on the same samples taken over time. This was done by Davies and Williams (1984) in an attempt to resolve the important question of whether the ^{14}C determination of photosynthesis agreed with photosynthesis as measured by the oxygen technique. Ecological data had previously indicated that results obtained by these two techniques could differ by a factor of 10. Within the precision of the two techniques, the authors found excellent agreement providing allowance was made for the fact that in the presence of nitrate, the photosynthetic quotient used to convert oxygen values to ^{14}C photosynthetic values is increased appreciably.

Another step in improving the methodology of measuring plankton dynamics has been achieved by Sonntag and Parslow (1981) and Hay et al. (1988). Because mesocosm experiments offer the best time-series data on zooplankton growth through various stages, the authors were able to specify a dynamic zooplankton population model which could be fitted by least squares analysis to the actual data. Thus it was possible to obtain population parameters and production estimates, as well as clarifying the data-set by smoothing and interpolation for purposes of analyzing and comparing different treatments with each mesocosm.

Testing of empirical models derived as statistical relationships between plants or animals and one or two environmental factors is a second use of mesocosm data. Doering and Oviatt (1986) compared five empirical models of bivalve filtration rates published by different authors. In their experiments, the amount of particulate material sedimented due to the feeding activity of the bivalve *Mercenaria mercenaria* in 13 m^3 mesocosms was compared with model results. The latter were based on temperature and/or size and had been derived from small-scale laboratory experiments using natural particles, algal monocultures, or dyes. Only those models which were based on filtration rates with natural particulate matter gave results agreeable with sedimentation rates. Models using dyes or monocultures tended to overestimate bivalve filtering rates.

In a similar approach to the above, Reynolds et al. (1983) established an empirical equation for the mean net rate of change (k_{nm}) of a population of *Fragilaria* as

$$k_{nm} = 0.292 (\log_{10} z_m) - 0.028$$

where z_m was the mixed layer depth. When this model was tested under conditions of variable water clarity, a second component had to be added for effective prediction, such that

$$k_{nm} = 0.292(\log_{10}z_m) - [1/\{1.048-0.039(z_m/z_s)^2\}] + 0.972$$

where z_s was the Secchi disk depth.

Whereas these examples of empirically-derived statistical relationships may have some limited value, particularly in the environment in which they were derived, the general tendency has been to derive ecosystem models based on deterministic relationships. From the large amount of physiological data on photosynthesis versus light intensity, zooplankton feeding relationships, temperature growth coefficients, etc., as well as environmental data on such parameters as the range of extinction coefficients, mixed layer depths, etc. in different oceans, it is possible to derive deterministic models in which each parameter is linked together by a set of differential equations. These models have generally been developed to a degree of sophistication required to answer a specific question. In some cases, therefore, the model need only be a caricature of events; in other cases, an attempt may be made to reproduce the major events in a mesocosm ecosystem.

Steele and Hendersen (1981) constructed a predator/prey ecosystem model of phytoplankton and zooplankton using one nutrient as the forcing function. The model is described as a caricature of an ecosystem rather than an attempt to synthesize total ecosystem function. As such, it was set up specifically to test the hypothesis that, on capture, enclosures contain a particular level of predators which tends to force other components to relatively high or low concentrations. In contrast, outside the enclosures, the same lower trophic level components are subject to highly variable (i.e. random) levels of predation, thus keeping the system from extremes in biomass production of any one trophic level. The authors compared the actual results of experiments in which plankton levels were measured outside enclosures as well as within. Some experimental results from the CEPEX and Loch Ewe programs (Fig. 1) tended to support the hypothesis of greater stability in plant and animal populations outside the enclosure. The mechanism causing this stability was investigated using the computer model. The authors compared the effect of predators and nutrient flux on the standing stock of phytoplankton and zooplankton. The effect of imposing initial high or low planktivore predation on the system was to drive the standing stocks to extremes. When the planktivore population was subjected to random variations, the fluctuations in phytoplankton and zooplankton were relatively small (Fig. 2). Random changes in the nutrient field were less significant in controlling fluctuation in the phytoplankton/herbivore standing stock. In conclusion, the authors emphasized that random processes may have an important function in ecosystem ecology, and these may be largely eliminated in studies conducted in enclosed ecosystems.

Randomness in food supply was also considered by Wangersky (1982) and Wangersky and Wangersky (1983). In the former reference, the author specifically considered the consequence of a random nutrient supply, such as from regenerative processes, on the diversity of organisms in a mesocosm. Using a mathematical model, the author argued that both species diversity and patchiness of plankton can be a function of random nutrient availability (i.e. the fortunate placement of species near a site of regenerated nutrients). The author concluded that this will lead to high variability between replicate

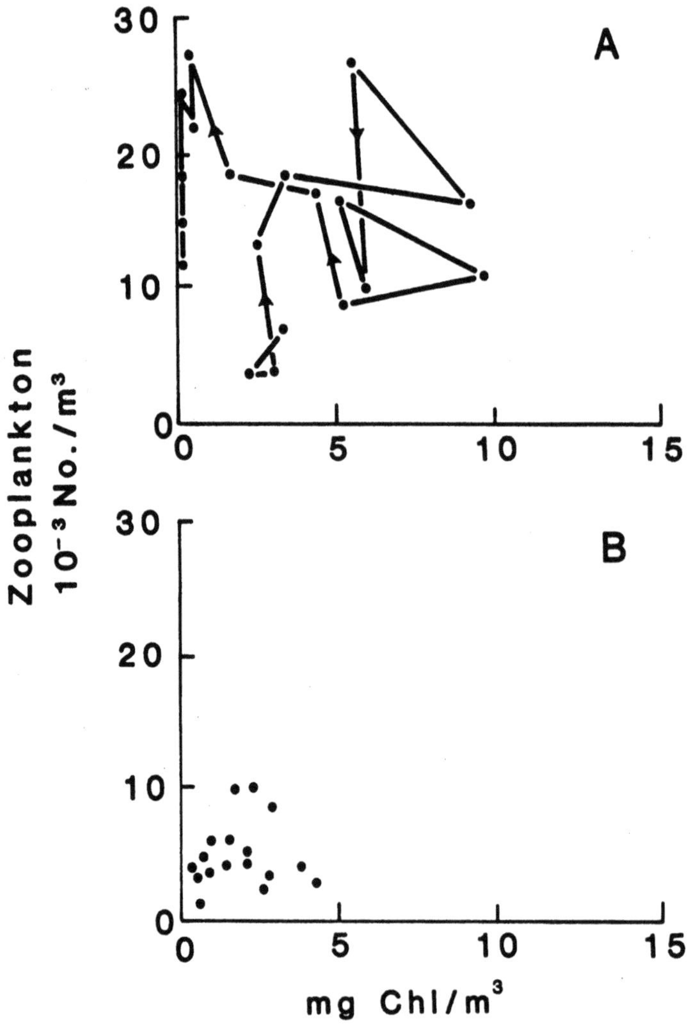

Figure 1. Phase plots of phytoplankton and herbivores in Loch Ewe mesocosms. A, Enclosure with herring larvae; B, water column outside the enclosure. From Steele and Hendersen (1981).

containers. The problem was further considered irreducible, because mixing of containers to eliminate the effect of nutrient patchiness could not be considered a valid simulation of the natural environment, except under conditions of violent turbulence or upwelling. However, in a comparison of nutrient-enriched versus control mesocosms, Parsons et al. (1977) did not find any significant difference in species diversity of the phytoplankton or protozoan community over a period of 30 days. During this period, diversity in both communities decreased, probably due to the isolation of the water columns. Whereas the total diversity of these two communities did not change, it is important to note that the nutrient treatment did cause a shift in the relative species composition of diatoms and microflagellates, which is predictable from nutrient uptake kinetics.

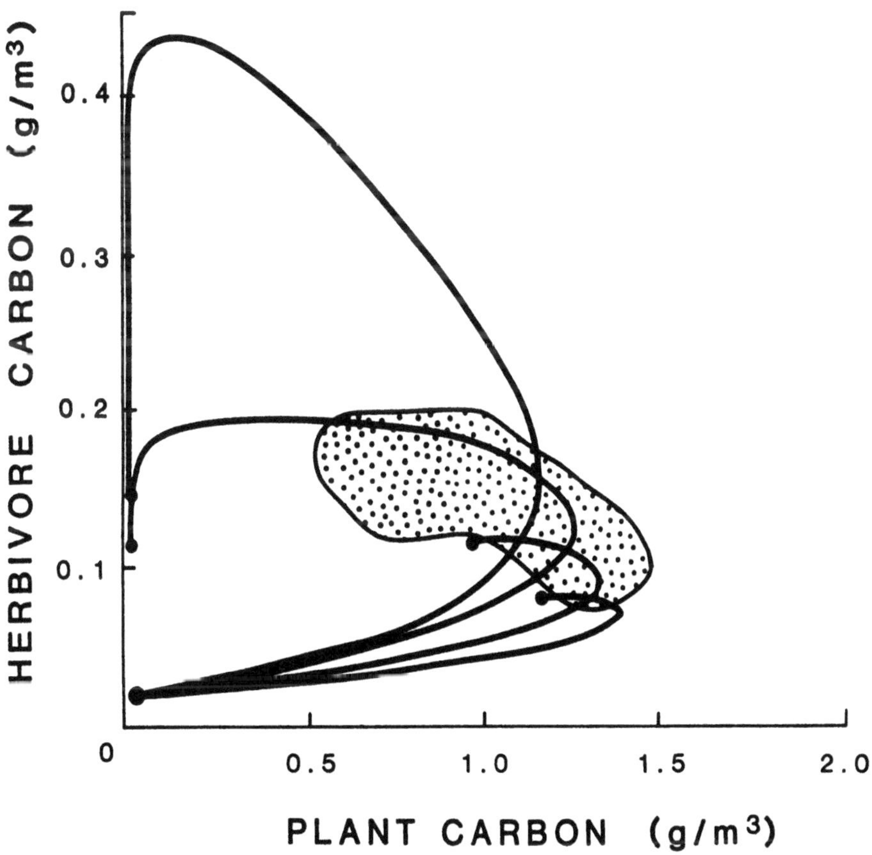

Figure 2. Phase plots of phytoplankton and herbivores derived from the Steele and Hendersen (1981) model. Four solid lines indicate phases generated in herbivore/phytoplankton populations when ctenophore predation levels are varied randomly over the range of values.

In general, this review does not include a discussion of models associated with chemostat cultures or microcosm batch culture experiments. This subject is well reviewed in specific references (e.g. Bazin, 1981). However, an important question related to mesocosm research is dealt with by Bratbak and Thingstad (1985). In this paper, an attempt was made to analyze the competition between algae and bacteria for a limiting inorganic nutrient. The results of chemostat cultures showed that, under conditions of phosphorus limitation, there was an increase in the bacterial to autotroph biomass ratio. This observation could be simulated using a simple mathematical model if it were assumed that nutrient-limited algae excreted dissolved organic carbon which acted as a substrate for bacterial growth. This theoretical explanation was later (Bratbak, 1987) verified experimentally using batch cultures of phytoplankton, bacteria, and bactivorous nanoflagellates.

A model of the bacterial loop was constructed by Laake et al. (1983) to account for the interplay of carbohydrates, bacteria, and zooflagellates in a 15-ton mesocosm. Four state variables defined by four equations were developed relating the quantities of bacteria, flagellates, ciliates, and organic substrate to each other; organic substrate was derived as a fractional release of the primary production and this was then consumed according to Michaelis-Menten kinetics by the bacteria. Growth of flagellates was dependent on the concentration of bacteria and their consumption by ciliates. Ciliate growth was considered to be both a function of the bacterial biomass and the flagellate biomass. Results obtained with this model were in good agreement with 20 days of observations within the mesocosms.

Three larger ecosystem models have been used to analyze events in marine mesocosms (Nixon et al., 1979; Parsons et al., 1986 and Parsons and Kessler, 1987; Andersen et al., 1987). These three models are essentially similar in that they all use time-dependent, coupled-process equations involving nonlinear empirical relationships and exogenous forcing functions, such as light and nutrient concentrations. Michaelis-Menten or Ivlev kinetics are used for nutrient and higher trophic level relationships. In spite of the general similarities in construction of the three models, there are also many differences. Some of these are summarized in Table 1.

The model by Nixon et al. (1979) was constructed to simulate the occurrence of winter blooms of phytoplankton in Narragansett Bay compared with adjacent mesocosms. The results showed that much more prolific growth of phytoplankton could be obtained in mesocosms than in the bay; this effect was analyzed using the model. I_{opt}, a physiological property of the algae and defined as the light intensity at maximum growth, was found to be much lower in the mesocosm community. By using a value of I_{opt} = 20 ly day^{-1}, model simulations closely matched observed results. The conclusion from these experiments was that models should be used not only to account

Table I.

A comparison of three ecosystem models designed for use with mesocosms.

PARAMETER	MODEL Nixon et al. (1979)	Andersen et al. (1987)	Parsons et al. (1986); Parsons & Kessler (1987)
PHYTOPLANKTON			
Standing stock components	2 species	Diatoms and flagellates	Diatoms and flagellates
Radiation	Derived from surface, k, etc.	Measured	Derived from surface, k, etc.
Nutrients	3 nutrients, 1 limiting growth	2 nutrients, 1 limiting growth	1 nutrient limiting growth
Temperature	Temperature control	Temperature control	Temperature control
HETEROTROPHIC CYCLE	None	None	Bacteria and zooflagellates
ZOOPLANKTON			
Standing stock components	Macro zooplankton	Copepods and larvaceans	Macro- and microzooplankton
Prey density, growth, and assimilation relationships	Yes	Yes	Yes
Threshold prey concentration	None	Yes	Yes
Temperature functions	Yes	Yes	Yes
Selective feeding	No	Diatoms or flagellates	Diatoms or flagellates
Zooplk. stages	Yes	No	No
PLANKTIVORES			
Fish	Yes	No	Yes
Ctenophores	Yes	Yes	Yes

for observations, but also in the design (especially regarding light conditions) of mesocosms.

The model described by Andersen et al. (1987) was designed to analyze ecological events in CEPEX mesocosms. In an initial model the authors used single compartments for phytoplankton, herbivores, and nutrients. This model did not provide an adequate explanation of ecological events within the ecosystems. A second model was constructed in which two nutrients were introduced together with a subdivision of phytoplankton into flagellates and diatoms, and zooplankton into larvaceans and copepods. Agreement between environmental data and model output was then found to be quite reasonable, except for carnivorous plankton (ctenophores) which were overestimated. An additional useful aspect of this model was the output of nitrogen data in different trophic compartments on different days.

The model by Parsons et al. (1986) was designed to account for changes in a plankton community following the addition of mine tailings to a mesocosm. Unlike the model described by Andersen et al. (1987), which was specifically designed to simulate events within a mesocosm, the model described by Parsons et al. was designed for more general use in diagnosing specific events known to occur in mesocosms and coastal environments. As an example of one of these events, it was found that mine tailings had the unexpected effect of increasing zooplankton production in a mesocosm. This effect could be simulated by introducing a range of extinction coefficients. Higher zooplankton production was then found to result from better phasing between phytoplankton and zooplankton production, at an optimal light intensity that slowed phytoplankton growth to a rate at which they were more efficiently consumed by zooplankton. More recent versions of the same model (Parsons and Kessler, 1987; Parsons and Taylor, in press) have been used to diagnose the effect of higher trophic levels and the effect of mixing in mesocosms. In the latter experiments, a comparison was made with results obtained in an earlier mesocosm experiment involving vertical mixing (Sonntag and Parsons, 1979). Using the model to diagnose mixing effects, it was found that sustained mixing (at a rate of 0.1 m h^{-1}) caused a high production of ctenophores. This result (Fig. 3B) was similar to that obtained under experimental conditions (Fig. 3A), although no attempt was made to replicate the scale of events because this depended on a large number of factors including the starting conditions and the intensity of upwelling, which was not known in the experiments. The higher production of ctenophores in the mixed water column was not simply a function of greater nutrient availability due to upwelling. A phase diagram of phytoplankton and zooplankton standing stock produced by the model (Fig. 3C) shows that the zero-mixed water column underwent much greater extremes in phytoplankton and zooplankton populations compared to the mixed water column. This indicates better

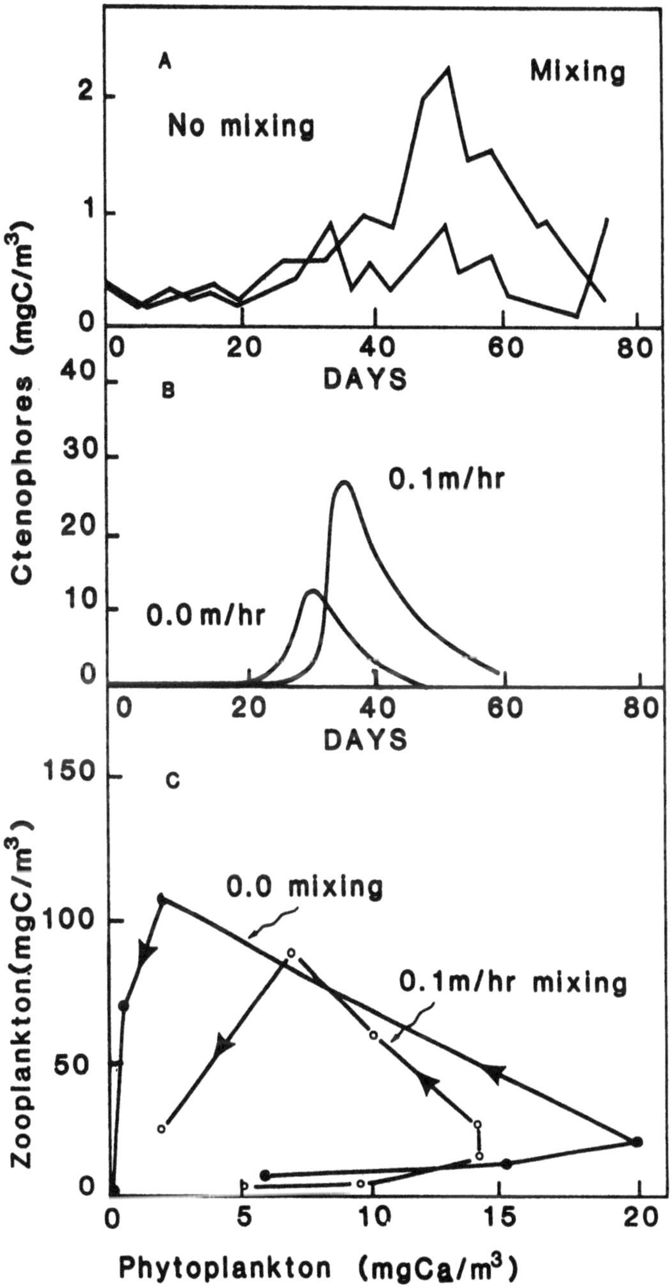

Figure 3. A comparison of the production of predators in upwelled ecosystems. A, Ctenophores (mixing curve) and *Sagitta* (no mixing curve); B, ctenophore model results ; C, zooplankton - phytoplankton phase diagram over 35 days, with and without mixing. A from Sonntag and Parsons (1979); B and C from Parsons and Taylor (in press).

phasing (or coupling) of primary and secondary production in the mixed water column. Excessive production of primary and secondary producers under conditions of no mixing leads to greater respiratory and sinking losses, with less transfer to higher trophic levels in the pelagic environment. Trophodynamic phasing of ecosystems is therefore an important property of their function and is related to the ecological efficiency of various components in the system.

The models described above are all strongly quantitative, simplistic in that the model structure is generally reduced to the least number of major pathways for energy flow, and deterministic in the sense that known physiological relationships derived from laboratory experiments are incorporated into the network of relationships. In a series of papers (e.g. Lane and Levins, 1977; Lane, 1986), it has been suggested that a more qualitative approach to aquatic ecosystems would better represent the large number of interconnecting links between the whole array of nutrients and species that are generally present in natural ecosystems. The analytical model employed in the papers cited above is known as "loop analysis", and it has been applied to the results of mesocosm research (Lane and Collins, 1985). The main advantage of the model is its ability to deal with a large number of components; the disadvantage seems to lie in the reduction of interactions to zero, positive, negative, or ambivalent (if, in the latter case, there are both positive and negative interactions on the same component). Models can be constructed to fit the data; in the case of the mesocosm experiments, the models employed were fitted with 95% agreement in predictions with data values. Perhaps one of the more useful outcomes from the models is the tendency to explain counter-intuitive results. For example, in a simple case, nutrients may not increase algal production of one species because of the presence of selective predation on the same species by a herbivore. With an extensive model network, such reactions can be observed and they serve as explanations for conflicting results which have been obtained in field experiments cited by the authors. For any one area, it appears that much of the annual cycle of an ecosystem can be represented by a "core" structure that responds throughout the year to different external forces (e.g. seasonal changes in light, etc.). In conclusion to the references on "loop analysis", the authors emphasize that their technique is not aimed at replacing quantitative models but is a means of obtaining more extensive explanations of nature.

CONCLUSIONS

From simple statistical comparisons of techniques to sophisticated models of ecosystem dynamics, the results of mesocosm experiments have been used to ground-truth empirical and physiological relationships established from environmental and laboratory experiments. The models have assisted in the understanding of ecosystems, and the ecosystem changes have assisted in modifying the models. This is a two-way scientific process that has been beneficial to experimentalists and theoreticians. Future attempts to model ecosystem processes could learn from the experience gained from mesocosm research in aquatic environments.

One important consideration in the comparison of mathematical ecosystem models with the results of ecosystem data collected from mesocosms or from the field, is that the mathematical model is a scenario of events that may explain results but may not be able to predict them. This distinction was clearly made by Bradbury et al. (1984), and an illustration from their paper is given in Fig. 4. In this illustration, the inherent unpredictability of all ecosystem models is emphasized on the grounds that nature is more infinitely resourceful than can be described by any ecosystem model that falls short of describing nature itself. Thus, in Fig. 4, the ecosystem model may be

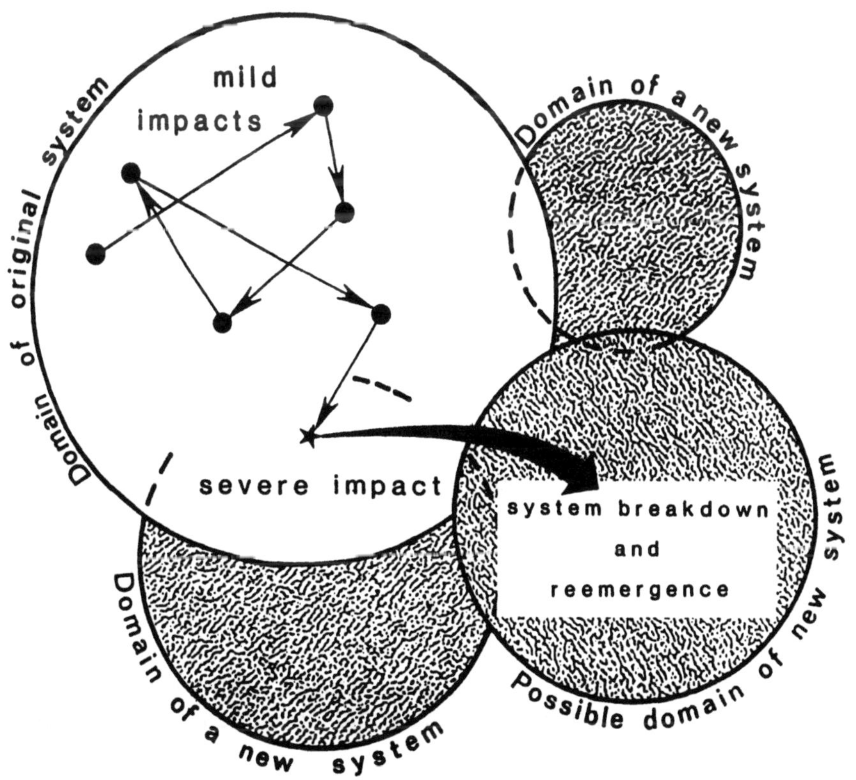

Figure 4. Possible trajectories of a perturbed ecosystem (modified from Bradbury et al., 1984). Each large circle may be analogous to "core" structures of varying complexity as described by Lane and Collins (1985).

sufficient to simulate a number of changes that could occur as a result of perturbations (climatic or pollutant) within the domain for which it was constructed. However, if a severe impact forces the ecosystem into some new domain, the ecosystem model is no longer a meaningful representation of nature. At this point, the model's representation of nature collapses. Conceptually, this appears to be analogous to the collapse of a "core" structure as described by Lane and Collins (1985) and its replacement by another "core" structure represented by a different model. For mesocosm research, where the ecosystem is already physically confined to the domain of a water column, mathematical models will tend to be more comparable with experimental data than with data collected from the field, where there is a much greater opportunity for change.

LITERATURE CITED

Andersen, V., P. Nival, and R. Harris. 1987. Modelling of a planktonic ecosystem in an enclosed water column. *J. Mar. Biol. Ass. U.K.* **67**: 407-430.

Bazin, M. J. 1981. Theory of continuous culture. Pp. 27-62. **In**: P. H. Calcott [ed.], *Continuous Culture of Cells*.. Boca Ratton, Florida: CRC Press.

Bradbury, R. H., L. S. Hammond, and R. E. Reichelt. 1984. Prediction versus explanation in environmental impact assessment. *Search* **14**: 323-325.

Bratbak, G. 1987. Carbon flow in an experimental microbial ecosystem. *Mar. Ecol. Prog. Ser.* **36**: 267-276.

Bratbak, G. and T. F. Thingstad. 1985. Phytoplankton-bacteria interactions: an apparent paradox? Analysis of a model system with both competition and commensalism. *Mar. Ecol. Prog. Ser.* **25**: 23-30.

Davies, J. M. and P. J. B. Williams. 1984. Verification of ^{14}C and O_2 derived primary organic production measurements using an enclosed ecosystem. *J. Plankton Res.* **6**: 457-474.

Doering, P. H. and C. A. Oviatt. 1986. Application of filtration rate models to field populations of bivalves: an assessment using experimental mesocosms. *Mar. Ecol. Prog. Ser.* **31**: 265-275.

Hay, S. J., G. T. Evans, and J. C. Gamble. 1988. Birth, growth and death rates for enclosed populations of calanoid copepods. *J. Plankton Res.* **10**: 431-454.

Laake, M., A. B. Dahle, K. Eberlein, and K. Rein. 1983. A modeling approach to the interplay of carbohydrates, bacteria and non-pigmented flagellates in a controlled ecosystem experiment with *Skeletonema costatum*. *Mar. Ecol. Prog. Ser.* **14**: 71-79.

Lane, P. A. 1986. Symmetry, change, perturbation, and observing mode in natural communities. *Ecology* **67**: 223-239.

Lane, P. A. and R. Levins. 1977. The dynamics of aquatic systems. 2. The effect of nutrient enrichment on model plankton communities. *Limnol. Oceanogr.* **22**: 454-471.

Lane, P. A. and T. M. Collins. 1985. Food web models of a marine plankton community network: An experimental mesocosm approach. *J. Exp. Mar. Biol. Ecol.* **94**: 41-70.

Nixon, S. W., C. A. Oviatt, J. N. Kremer, and K. Perez. 1979. The use of numerical models and laboratory microcosms in estuarine ecosystem analysis - Simulation of a winter phytoplankton bloom. Pp. 165-189. **In**: R. F. Dame [ed.], *Marsh-Estuary Systems Simulation*. Columbia, SC: Univ. South Carolina Press.

Parsons, T. R., W. H. Thomas, D. Seibert, J. R. Beers, P. Gillespie, and C. Bawden. 1977. The effect of nutrient enrichment on the plankton community in enclosed water columns. *Int. Rev. Ges. Hydrobiol.* **62**: 565-572.

Parsons, T. R., T. A. Kessler, and Li Guanguo. 1986. An ecosystem model analysis of the effect of mine tailings on the euphotic zone of a pelagic ecosystem. *Acta Oceanologica Sinica* **5**: 425-436.

Parsons, T. R. and T. A. Kessler. 1987. An ecosystem model for the assessment of plankton production in relation to the survival of young fish. *J. Plankton Res.* **9**: 125-137.

Parsons, T. R. and A. H. Taylor. (In press) The effect of vertical mixing on ecosystem dynamics in large mesocosms. Proceedings IMEEES Symposium Beijing, 1987.

Platt, T., K. H. Mann, and R. E. Ulanowicz. 1981. *Mathematical Models in Biological Oceanography. UNESCO Monograph on Oceanic Methodology*, No. 7. Paris: The UNESCO Press. 156 pp.

Reynolds, C. S., S. W. Wiseman, B. M. Godfrey, and C. Butterwick. 1983. Some effects of artificial mixing on the dynamics of phytoplankton populations in large limnetic enclosures. *J. Plankton Res.* **5**: 203-234.

Sonntag, N. C. and T. R. Parsons. 1979. Mixing an enclosed, 1,300 m^3 water column: effects on the planktonic foodweb. *J. Plankton Res.* **1**: 85-102.

Sonntag, N. C. and J. Parslow. 1981. Technique of systems identification applied to estimating copepod production. *J. Plankton Res.* **3**: 461-473.

Steele, J. H. and E. W. Hendersen. 1981. A simple plankton model. *Am. Naturalist* **117**: 676-691.

Wangersky, P. J. 1982. Model ecosystems: the limits of predictability. *Thalass. Jugoslavica* **18**: 1-10.

Wangersky, P. J. and C. P. Wangersky. 1983. The Manna effect: paradox of the plankton. *Int. Rev. Ges. Hydrobiol.* **68**: 327-338.

SUBJECT INDEX

Numbers in bold italics refer to pages with text-figures

Acartia tonsa, 51, 73-74
Achnanthes taeniata, 173
Actinomonas, 12
Aggregates, 17-***28***
Amphipods, 49-51, ***50***
Amphiprora paludosa, 49
Anaerobic microcosms, 37-39
Anisogammarus pugettensis, 49
Arenicola, 114
Arkona Sea, *see* Baltic Sea
Artemia salina, 33
Ascophyllum nodosum, 128
Asterias rubens, 129
Asterionella formosa, 68, ***70***
Atlantic Ocean, 68, 159

Bacteria, ***14***, 23-24, 30, 32, 34, 48, 51, 52, 74, 83, 85, 93, 161, 202.
See also Cyanobacteria; Sulfur bacteria
Bacterioplankton, 12-14, ***13***
Baltic Sea, 110, 123, 124, 164, 169-185
Barents Sea, 148
Benthic macrofauna, 37-41, 42
Benthic microalgae, 39
Benthic primary production, 39
Biofouling, 32, ***33***, 62
Biological oceanography
research problems, 2, 4, 82-83
Brachionus, 49
Branchiomonas submarina, 48
Bremerhaven caissons, 111, 163
Bryozoa, 113
Buoyancy, 89

Cachonia niei, 67
Calanus, 87-***88***
  finmarchicus, 13
Capelin, 147-148, 150
Carbon
  flux, 103, 112
  particulate organic, 37
Carcinus maenas, 40, 129
Ceratium furca, 71
Chaetoceros, 173, 182, 183
  constrictus, 89
Chemostat, 21-30, ***22, 26***, 48, ***49***
Cholera, 52
Chrysochromulina polylepis, 97
Ciliates, 12-13, ***14***, 48, 52, 202
Cladocera, ***72***
Cladophora rupestris, 129
Cod, ***141***, ***142***, 145, 147, 148, 150
Competition, 22-25, ***23***, 34, 202
Computer models, *see* Models, computer
Continuous culture, 22-32, ***27***, ***50***
Copepods, 13-***14***, 42, 43, 51, ***65***, 67, ***69***, ***72***, 73, 112, 204
Coral reef mesocosms, 126, ***127***, 128, 130-131
Corophium volutator, 39
Coscinodiscus radiatus, 176
Costs, relative, ***8***, 117
Crassostrea gigas, 40
Ctenophores, 63, 67, 73-***75***, ***201***, 204, ***205***
Currents, water, 131-132
Cyanobacteria, 12-***14***, 175-176, 182-184
Cyclotella, 71

Daphnia, 91
  magna, 49
Decomposition, 87, 93
Design criteria, 5
Detritus, 37, ***38***
Diatoma, 71
Diatoms, 64-67, 71, 73, 74, 87, 89, 91, 97, 100, 175-176, 182-184, 201, 204
Dinoflagellates, 67, 71, 91, 97, 100, 173, 175-176, 182-184

Dissolved organics
    release by phytoplankton, 85-***86***, ***92***-93, ***94***, ***95***
Distribution, in enclosures, 189-190, 200-201
Ditylum brightwellii, 67
Diurnal rhythms, 82, 92-93, 130
Diversity, of species, 29, 63, 129, 170-171, 200-201
Duration time, 1, 10, 62, 64, 81, 111, 189

Ecological efficiency, 15, 73-74
Ecosystem
    definition, 11
Estuarine simulation, 27-29, 30-32, 39-40, 110-113
Euphausiids, 64, 147
Euphotic zone, 68, 71
Eurytemora
    cholerae, 51
    hirundoides, 43, ***45***
Eutrophic water, 9, 71
Eutrophication, ***44***, ***47***, 82, 84, 112, 169-185
Excretion
    by macrofauna, 40-41
Experimental ecosystem
    definition, 1
    uses, 3

Fatty acids, 150
Fecal pellets, 13, ***14***, 113
Feeding experiments, 48-51, 198
Femtoplankton, ***13***
Fish, 132, 136-150. See also Capelin; Cod; Herring; Plaice; Salmonids; Turbot
Fisheries
    production, 68
    research problems, 4, 136-150
Flagellates, 12, ***14***, 34, 64-67, 74, 83, 88, 91, 201, 202, 204
Flow cytometry, 21
Fluxes, ecological, 2, 21, 39, 40-41, 82, 96, 100, 102
Food chains, 12, 15, 61-76
Food partitioning, 51
Food webs, 12, 14, 46, 48, 71, 73
Fragilaria, 198
Free-floating systems, 6, 76, 111

Fucus
distichus edentatus, 129
vesiculosus, 122-***125***, 126
Fungi, 68, ***70***

Gel-stabilized model systems, 30-32
Generation time, ***8***, 62, 64, 81
Gnotobiotic
definition, 11
experiments, 10, 11, 21, 33
Gonyaulax polyedra, 67
Gradostat, 21, 22, ***29***, 30, ***31***
Grazing, 26, 64, 67, 82, 87-***88***, 100, 133
Growth
barrier, 143, 146-147
efficiency, 48
phasing, *see* Temporal phasing
rate, 22-***25***, ***23***, ***24***, 48, 143-147, ***144***
Gyrodinium aureolum, 91, 97, 100

Heat dissipation, ***44***, ***45***, ***47***
Herring, 63, ***144***, 145, 147, 148, 150
Heterosigma akashiwo, 91
Hydrobia, 39, 42
ventrosa, 44, ***45***

Isotopes
use of in mesocosms, 71, 73, 102-103

Jellyfish, *see* Medusae

Karlskrona, 122-***125***, 126, 128-133

Laminaria
digitata, 129
saccharina, 126, 129
Larvacea, ***69***, 204
Leptocylindrus minimus, 100
Littorina littorea, 129
Lohmanniella, ***65***
Loop analysis, 206

Macrocosms, 140, 143-150
    definition, 1, 2, 137
Macroplankton, ***13***
Marine Ecosystem Research Laboratory, *see* MERL
Marine snow, ***14***
Medusae, 147, 149
Megaplankton, ***13***
Meiofauna, 42, 43, 51, 113
Mercenaria mercenaria, 103, 198
MERL, 10, 35, 103, 109-113, 116, 158-162, 191
Mesocosms, 140, 142-143, 148, 155-166, 169-185, 188-194, 197-208
    benthic, 1, 109-117, 122-134
    definition, 1, 2, 7-8, 20, 137, 157
    pelagic, 61-76, 81-103, 170-185
Mesodinium rubrum, ***65***
Mesoplankton, ***13***
Metabolic rates, 21
Metazooplankton, ***13***, 14
Microbial films, 32, ***33***
Microcalorimetry, 41-46, ***43***
Microcosms, 20-53, ***36***, ***50***. *See also* Sediment microcosms
    benthic, 1, 37-46, ***41***
    definition, 1, 20
Microplankton, 12, ***13***, 14
Mine tailings, 68-***69***, 204
Mineralization, 37-40
Models, computer, 82, 197-208
Monod growth kinetics, ***23***
MODUS, 110
MOTIF, 110
Mycoplankton, ***13***
Mytilus edulis, 129

Nanoplankton, ***13***, 14, 27
Narragansett Bay, 35, 110, 111-112, 159, 202
Nekton, ***13***, 15, 63
Nematodes, 51
Nereis, 114
    virens, 37-39, ***38***
Netherlands Institute for Sea Research, *see* Texel

Nitrogen
decay rates, 37-40
detrital, 37
flux, 100, 102
Nutrient
addition, *see* Eutrophication
flux, 112
limitation, 22-25, 82, 96, 202
partitioning, 25
regeneration, 26, 102
uptake, 22-25, 82, 181-182

Oceanization, 172
Oil, 111, 126, 133, 164
Oligotrophic water, 9, 27

Pacific Ocean, 25, 68
Paracalanus, 97, ***99***
parvus, 73-74
Particle-size distribution, 11-12, ***101***
Pavlova lutheri, 48-***49***
Pelagic ecosystems, 11-15
Phaeocystis, 67
Phasing, *see* Temporal phasing
Phymatolithon lenormandii, 129
Phytoplankton
experiments, 10, 15, 22, 25, 34-36, 63-75, 85-97, 172-185, 202
Picoplankton, 12, ***13***
Plaice, 147
Pleurobrachia
bachei, 73-74
pileus, 100
Point of no return (PNR), 146
Pollutants
experiments, 111, 112, 126, 159-164, 192
fates and effects, 2, 74, 100, 157-166
oil, 111, 126, 133, 164
organic compounds, 158-162
metals, 163
research problems, 5, 82, 84, 155-166
Predator-prey relations, 132-133, 137, 140, 147-150, 199-***200***, ***201***
Production, biological, 34, 39, 68, 71, 73, ***75***, 85, 114, 171-185

Protozoa, ***14***
Protozooplankton, 12, ***13***, ***14***
Pseudocalanus, 87-***88***
Pseudomonas, ***24***
Pyramid of numbers, 12

Red tide, 91
Redox potential, 30, ***31***
Remineralization rate, 12
Replication, 20, 63-64, 76, 111, 114, 122, 128-130, 137, 140, 165, 189-191, 194, 200-201
Reproducibility, 63-64, 84, 130-132, 137, 156, 190
Resources
    exploitation, 2, 4
    management, 2, 4
Rhizosolenia fragillissima, 176
Rhodomonas minuta, 176
Rotifers, 48-49, ***72***

Sagitta, ***205***
Salmonids, 63
Sediment microcosms, 30-32, 37-46, ***41***. *See also* Mesocosms, benthic
Sedimentation, 87-90, 198
Seston, ***14***
Sewage, 111
Silicate limitation, 97
Single-species experiments, 5, 10
Sinking, *see* Sedimentation
Size
    of enclosures, 1, ***8***, 9, 20, 33-34, 40, 189
    of food, 12
    of organisms, ***8***, 12, ***13***, 100-***101***, 171
Skeletonema costatum, 25, 73, 96-97, ***98***, 173, 176, 182, 183
Solar radiation
    reduction of, 68-***69***, 71-***72***, 142
Solbergstrand, 110, 113, 122-126, ***125***, 128-133
Spartina, 37
Spirillum, ***24***
Starvation, 145, 147, 149, 184
Statistical treatment, 188-194
Steady state, 22-25, 27
Storm effects, 112

Strombidium, 48-***49***, ***65***
Sublittoral benthic systems, 6
Succession, 25, 34, 35-36, 46, 64-67, 82, 88, 91, 93, 95-100, 129
Sulfur bacteria, 24, ***25***, 30, ***31***, 32
Synechococcus, 12-13
Synthetic microcosm, 11, 33

Tay Estuary, 30, 32
Temporal phasing, 67-71, 204-206
Texel, 110, 114-115
Thalassiosira, 182, 183
    pseudomonas, 25
    rotula, 85-***86***, ***92***, ***94***, ***95***, 96, ***98***, 173
Thermodynamics, 46
Trinity Bay, 28
Trophic levels, 9, 11-12, 73-74, ***75***, 82, 100
Turbot, 145, ***146***, 147
Turbellaria, 51
Turbulence, 34-36, 40-41, 87-88, 91, 131-132, 149-150, 201. *See also* Vertical mixing

Upwelling simulation, 34, 63, 204
Uronema, 13, ***65***

Vertical migration, 82, 91
Vertical mixing, 68, 91, 143, 204
Vibrio cholerae, 51-52
Virioplankton, ***13***

Wadden Sea, 110-111

Zooplankton, 63, 71, 74
    experiments, 68-***69***, 87-***88***, 97, ***99***

Coastal and Estuarine Studies

(formerly Lecture Notes on Coastal and Estuarine Studies)

Vol. 1: J. Sündermann, K.-P. Holz (Eds.), Mathematical Modelling of Estuarine Physics. Proceedings, 1978. 265 pages. 1980.

Vol. 2: D.P. Finn, Managing the Ocean Resources of the United States: The Role of the Federal Marine Sanctuaries Program. 193 pages. 1982.

Vol. 3: M. Tomczak Jr., W. Cuff (Eds.), Synthesis and Modelling of Intermittent Estuaries. 302 pages. 1983.

Vol. 4: H.R. Gordon, A.Y. Morel, Remote Assessment of Ocean Color for Interpretation of Satellite Visible Imagery. 114 pages. 1983.

Vol. 5: D.C.L. Lam, C.R. Murthy, R.B. Simpson, Effluent Transport and Diffusion Models for the Coastal Zone. 168 pages. 1984.

Vol. 6: M.J. Kennish, R.A. Lutz (Eds.), Ecology of Barnegat Bay, New Jersey. 396 pages. 1984.

Vol. 7: W.R. Edeson, J.-F. Pulvenis, The Legal Regime of Fisheries in the Caribbean Region. 204 pages. 1983.

Vol. 8: O. Holm-Hansen, L. Bolis, R. Gilles (Eds.), Marine Phytoplankton and Productivity. 175 pages. 1984.

Vol. 9: A. Pequeux, R. Gilles, L. Bolis (Eds.), Osmoregulation in Estuarine and Marine Animals. 221 pages. 1984.

Vol. 10: J.L. McHugh, Fishery Management. 207 pages. 1984.

Vol. 11: J.D. Davis, D. Merriman (Eds.), Observations on the Ecology and Biology of Western Cape Cod Bay, Massachusetts. 289 pages. 1984.

Vol. 12: P.P.G. Dyke, A.O. Moscardini, E.H. Robson (Eds.), Offshore and Coastal Modelling. 399 pages. 1985.

Vol. 13: J. Rumohr, E. Walger, B. Zeitzschel (Eds.), Seawater-Sediment Interactions in Coastal Waters. An Interdisciplinary Approach. 338 pages. 1987.

Vol. 14: A.J. Mehta (Ed.), Estuarine Cohesive Sediment Dynamics. 473 pages. 1986.

Vol. 15: R.W. Eppley (Ed.), Plankton Dynamics of the Southern California Bight. 373 pages. 1986.

Vol. 16: J. van de Kreeke (Ed.), Physics of Shallow Estuaries and Bays. 280 pages. 1986.

Vol. 17: M.J. Bowman, C.M. Yentsch, W.T. Peterson (Eds.), Tidal Mixing and Plankton Dynamics. 502 pages. 1986.

Vol. 18: F. Bo Pedersen, Environmental Hydraulics: Stratified Flows. 278 pages. 1986.

Vol. 19: K.N. Fedorov, The Physical Nature and Structure of Oceanic Fronts. 333 pages. 1986.

Vol. 20: A. Rieser, J. Spiller, D. VanderZwaag (Eds.), Environmental Decisionmaking in a Transboundary Region. 209 pages. 1986.

Vol. 21: Th. Stocker, K. Hutter, Topographic Waves in Channels and Lakes on the f-Plane. 176 pages. 1987.

Vol. 22: B.-O. Jansson (Ed.), Coastal Offshore Ecosystem Interactions. 367 pages. 1988.

Vol. 23: K. Heck, Jr. (Ed.), Ecological Studies in the Middle Reach of Chesapeake Bay. 287 pages. 1987.

Vol. 24: D.G. Shaw, M.J. Hameedi (Eds.), Environmental Studies in Port Valdez, Alaska. 423 pages. 1988.

Vol. 25: C.M. Yentsch, F.C. Mague, P.K. Horan (Eds.), Immunochemical Approaches to Coastal, Estuarine and Oceanographic Questions. 399 pages. 1988.

Vol. 26: E.H. Schumann (Ed.), Coastal Ocean Studies off Natal, South Africa. 271 pages. 1988.

Vol. 27: E. Gold (Ed.), A Law of the Sea for the Caribbean: An Examination of Marine Law and Policy Issues in the Lesser Antilles. 507 pages. 1988.

Vol. 28: W.S. Wooster (Ed.), Fishery Science and Management. 339 pages. 1988.

Vol.29: D.G. Aubrey, L. Weishar (Eds.), Hydrodynamics and Sediment Dynamics of Tidal Inlets. 456 pages. 1988.

Vol. 30: P.B. Crean, T.S. Murty, J.A. Stronach, Mathematical Modelling of Tides and Estuarine Circulation. 471 pages. 1988.

Vol. 31: G. Lopez, G. Taghon, J. Levinton (Eds.), Ecology of Marine Deposit Feeders. 322 pages. 1989.

Vol. 32: F. Wulff, J.G. Field, K.H. Mann (Eds.), Network Analysis in Marine Ecology. 284 pages. 1989.

Vol. 33: M.L. Khandekar, Operational Analysis and Prediction of Ocean Wind Waves. 214 pages. 1989.

Vol. 34: S.J. Neshyba, Ch.N.K. Mooers, R.L. Smith, R.T. Barber (Eds.), Poleward Flows Along Eastern Ocean Boundaries. 374 pages. 1989.

Vol. 35: E.M. Cosper, V.M. Bricelj, E.J. Carpenter (Eds.), Novel Phytoplankton Blooms. 799 pages. 1989.

Vol. 36: W. Michaelis (Ed.), Estuarine Water Quality Management. 478 pages. 1990.

Vol. 37: C.M. Lalli (Ed.), Enclosed Experimental Marine Ecosystems: A Review and Recommendations. X, 218 pages. 1990.

MIX
Papier aus verantwortungsvollen Quellen
Paper from responsible sources
FSC® C105338

If you have any concerns about our products,
you can contact us on
ProductSafety@springernature.com

In case Publisher is established outside the EU,
the EU authorized representative is:
Springer Nature Customer Service Center GmbH
Europaplatz 3, 69115 Heidelberg, Germany

Printed by Libri Plureos GmbH
in Hamburg, Germany